Thin Film Coatings

This book provides a thorough overview of the fundamental principles, design strategies, fabrication techniques, characterization methods, and diverse applications of thin films and advancements therein. With a balance between theoretical concepts, practical applications, and computational techniques, the book offers a holistic understanding of the subject matter, paving the way for students, researchers, and engineers to navigate the fabrication and characterization of thin films backed by real-world examples.

- Covers optimization of coatings tailored to specific applications, coupled with their industry-focused approach that incorporates real-world examples
- Discusses thin films in detail, including a thorough examination of deposition and analysis methods
- Explores emerging topics related to the endurance of coatings, highlighting green technologies and life cycle assessment
- Reviews the incorporation of machine learning and evolutionary algorithms for thin film optimization
- Delves into current innovations, recent models, and substantial research, with a focus on practical use cases

This book is aimed at researchers and graduate students in materials science, characterization, and thin films.

Advanced Materials Processing and Manufacturing

Series Editor: Kapil Gupta

The CRC Press Series in Advanced Materials Processing and Manufacturing covers the complete spectrum of materials and manufacturing technology, including fundamental principles, theoretical background, and advancements. Considering the accelerated importance of advances for producing quality products for a wide range of applications, the titles in this series reflect the state-of-the-art in understanding and engineering the materials processing and manufacturing operations. Technological advancements for enhancement of product quality, process productivity, and sustainability, are on special focus including processing for all materials and novel processes. This series aims to foster knowledge enrichment on conventional and modern machining processes. Micro-manufacturing technologies such as micro-machining, micro-forming, and micro-joining, and Hybrid manufacturing, additive manufacturing, near net shape manufacturing, and ultra-precision finishing techniques are also covered.

Manufacturing in the Digital Age
Edited by Mohan Kumar Pradhan, Mahendra U. Gaikwad, and Ragavanantham Shanmugam

Fused Deposition Modelling
Technology, Applications, and Developments
Edited by Anshuman Kumar Sahu and Siba Sankar Mahapatra

Manufacturing Tribology
Edited by Jibin T Philip, Chacko Preno Koshy, and MD Mathew

Machine Learning for Advanced Manufacturing
Edited by Nishant Ranjan, Rashi Tyagi, Ranvijay Kumar, Ashutosh Tripathi, and Amit Verma

Engineered Ceramics and Composite Materials
From Fundamentals to Applications
Edited by Neethu Bhaskar, Sathish Kumar Palaniappan, Sanjay Mavinkere Rangappa, and Suchart Siengchin

Thin Film Coatings
Fundamentals and Advancements
Edited by Lokeswar Patnaik, Sunil Kumar, Magdalena Łępicka, and Saikat Ranjan Maity

For more information about this series, please visit: www.routledge.com/Advanced-Materials-Processing-and-Manufacturing/book-series/CRCAMPM

Thin Film Coatings
Fundamentals and Advancements

Edited by
Lokeswar Patnaik, Sunil Kumar, Magdalena Łępicka, and Saikat Ranjan Maity

CRC Press
Taylor & Francis Group
Boca Raton London New York

CRC Press is an imprint of the
Taylor & Francis Group, an informa business

First edition published 2026
by CRC Press
2385 NW Executive Center Drive, Suite 320, Boca Raton FL 33431

and by CRC Press
4 Park Square, Milton Park, Abingdon, Oxon, OX14 4RN

CRC Press is an imprint of Taylor & Francis Group, LLC

ISBN: 978-1-032-80450-7 (hbk)
ISBN: 978-1-041-26511-5 (pbk)
ISBN: 978-1-003-75065-9 (ebk)

DOI: 10.1201/9781003750659

Typeset in Times
By Apex CoVantage, LLC

Contents

Editor Biographies

Lokeswar Patnaik is an assistant professor in the Department of Mechanical and Manufacturing Engineering at the National Institute of Advanced Manufacturing Technology (NIAMT), Ranchi. He obtained his PhD in mechanical engineering from the National Institute of Technology, Silchar, where his research focused on advanced materials and tribology, which was followed by an ME from Osmania University and a BTech from NIT Silchar. His research interests lie in surface engineering, thin films, mechanical characterization, and the tribological performance of materials. He has published widely in peer-reviewed journals and also secured several patents on innovative machine designs, polishing attachments, and material testing devices. His expertise bridges fundamental research with practical applications, advancing both industrial and academic knowledge. Moreover, he contributes to the academic community through editorial roles and has served as a guest editor for reputed international journals in the field of materials science and engineering.

Sunil Kumar is an assistant professor (Senior Grade) in the Department of Mechanical Engineering at the Amrita School of Engineering, Amrita Vishwa Vidyapeetham, Chennai campus. He holds a Diploma, BTech, ME, and PhD in mechanical engineering and has cultivated academic and professional experience through both teaching and industry engagements. His research interests include surface modification, thin films, tribology, additive manufacturing of tool steels, machining of hard materials, as well as advanced methodologies such as multi-criteria decision-making, optimization, and fuzzy systems. He has published extensively in reputed journals, contributed book chapters, and actively reviews for international scientific journals. Dr Kumar is an inventor, holding multiple Indian patents that reflect his focus on practical innovation in mechanical engineering and manufacturing technologies. His academic journey combines strong technical expertise with interdisciplinary approaches, advancing both research and teaching. At Amrita Vishwa Vidyapeetham, he continues to mentor students, contribute to collaborative projects, and drive forward cutting-edge research in mechanical and manufacturing engineering.

Magdalena Łępicka is an assistant professor in the Department of Materials and Production Engineering in the Faculty of Mechanical Engineering at Białystok University of Technology. She completed her postdoctoral fellowship in the Nanodevices Research Group at the International Iberian Nanotechnology Laboratory in Braga, Portugal, before advancing through teaching and research roles at her current institution. Her research focuses on tribology, surface engineering, and the performance of advanced materials, with applications spanning composites, coatings, and dental biomaterials. She has authored numerous peer-reviewed publications on topics such as functional fillers in glass-ionomer cements and the effects of hydrothermal fatigue on restorative dental materials.

In recognition of her innovative work, Dr Łępicka received nearly PLN 1.3 million in funding from the National Science Centre under the prestigious SONATA 20 competition to advance her research. Beyond her academic contributions, she serves as a topic and guest editor for various reputed journals. Passionate about materials innovation and interdisciplinary applications, Dr Łępicka continues to make significant contributions to both mechanical and biomedical engineering.

Saikat Ranjan Maity is an associate professor in the Department of Mechanical Engineering at the National Institute of Technology (NIT) Jamshedpur and associated with NIT Silchar. He holds a PhD in engineering from Jadavpur University, along with an MTech in materials engineering and a BTech in production engineering. Dr Maity's academic and research career spans advanced manufacturing processes, industrial engineering, materials science, optimization, and the application of decision-making and soft computing techniques in engineering problems. He has published extensively in reputed international journals, conferences, and edited volumes, contributing significantly to the field of mechanical and production engineering. Before joining NIT Silchar, he worked as a lecturer and later assistant professor at the Haldia Institute of Technology and also served as a research scientist on a collaborative project sponsored by the Ministry of Steel and TATA Steel, focusing on fatigue behaviour of welded and bonded high-strength steel joints. Committed to both teaching and research, Dr Maity is actively engaged in guiding students, developing interdisciplinary projects, and contributing as a reviewer and editor for academic publications. His work reflects a strong integration of theoretical knowledge with applied engineering practices.

Contributors

Tushar Banerjee
Department of Production and Industrial Engineering
National Institute of Technology
Jamshedpur, Jharkhand, India

Himanshu Batham
Department of Production and Industrial Engineering
National Institute of Technology
Jamshedpur, Jharkhand, India

Sukanti Behera
Department of Mechanical Engineering
Amrita School of Engineering
Amrita Vishwa Vidyapeetham, Chennai, India

Alfa Bisoi
Department of Mechanical Engineering
National Institute of Technology
Silchar, Cachar, India

Abhik Bose
Department of Production and Industrial Engineering
National Institute of Technology
Jamshedpur, Jharkhand, India

Venkata Sushma Chinta
Department of Mechanical Engineering
Chaitanya Bharathi Institute of Technology
Hyderabad, India

Amartya Shekhar Das
TUM School of Engineering and Design
Technical University of Munich
Arcisstraße 21, München, Germany

Jatin Gajjar
TUM School of Engineering and Design
Technical University of Munich
Arcisstraße 21, München, Germany

Alisha Kakkar
Department of Chemistry
Banasthali Vidyapith, Tonk
Rajasthan, India

Simanchal Kar
Department of Mechanical Engineering
National Institute of Technology
Silchar, Cachar, India

Amireddy Kiran Kumar
Department of Mechanical Engineering
Chaitanya Bharathi Institute of Technology
Hyderabad, India

Nishant Kumar
Department of Mechanical Engineering
Chaitanya Bharathi Institute of Technology
Hyderabad, India

Sudesh Kumar
Department of Chemistry, DESM
NIE, National Council of Educational Research and Training
New Delhi, India

Sunil Kumar
Department of Mechanical Engineering
School of Engineering, Amrita Vishwa Vidyapeetham, Chennai Campus
Tiruvallur, Tamil Nadu, India

Magdalena Łępicka
Faculty of Mechanical Engineering
Bialystok University of Technology
Wiejska 45C, 15–351, Bialystok, Poland

Adeeba Mirza
Department of Chemistry
Graphic Era (Deemed to be) University
Clement Town, Dehradun, Uttarakhand, India

Abhilasha Mishra
Department of Chemistry
Graphic Era (Deemed to be) University
Clement Town, Dehradun, Uttarakhand, India

Krishnadas Narayanan Nampoothiri
Department of Chemistry
Maulana Azad National Institute of Technology
Bhopal, Madhya Pradesh, India

Aswathy M. Narayanan
Department of Chemistry, Vel Tech Rangarajan
Dr. Sagunthala R&D Institute of Science and Technology
Vel Nagar, Avadi, Chennai, Tamil Nadu, India

Lokeswar Patnaik
Department of Mechanical Engineering
National Institute of Advanced Manufacturing Technology (Formerly NIFFT)
Ranchi, Jharkhand, India
Center of Excellence in Advanced Manufacturing Research, Innovation and Training [CoE-AMRIT]
National Institute of Advanced Manufacturing Technology (Formerly NIFFT)
Ranchi, Jharkhand, India

Yadav Narendra Kumar Rajbahadur
Department of Mechanical Engineering
National Institute of Technology
Silchar, Cachar, India

Avala Ramesh
Department of Chemistry
Maulana Azad National Institute of Technology
Bhopal, Madhya Pradesh, India

Ashutosh Sahu
Department of Mechanical Engineering
Chaitanya Bharathi Institute of Technology
Hyderabad, India

Manigandan Sekar
Division of Engineering
New York University
Abu Dhabi

Lavish Kumar Singh
School of Engineering
Jawaharlal Nehru University
New Delhi, India

V. S. S. Venkatesh
School of Mechanical Sciences
Indian Institute of Technology
Bhubaneswar, Odisha, India

Preface

The field of thin film coatings has developed out of a very practical need. In most engineering materials, the properties required in the bulk and at the surface are not the same. Strength, resilience, and toughness are demanded inside the material, while the surface must resist wear, reduce friction, and withstand corrosion. No single bulk material can satisfy all these requirements. The idea of depositing a thin coating on a suitable substrate made it possible to combine the two sets of properties. This is the starting point of thin film technology, and it continues to shape its growth today.

The early technologies of vapour-phase and liquid-phase deposition remain central. Electroplating, spin coating, chemical vapour deposition, and magnetron sputtering are well established; yet each has limitations in terms of control, adhesion, or scalability. These methods are the building blocks, and they appear at the beginning of this volume to provide a base for the reader. The choice of method is never neutral; it defines the type of film that can be achieved and the performance that can be expected.

From that foundation, attention moves to more advanced methods. Pulsed laser deposition, vacuum thermal evaporation, and other refined techniques have enabled industries such as microelectronics, optics, and medical devices to rely on thin films in everyday applications. The thin film in a semiconductor device or the protective coating in an implant is inseparable from the method by which it was created. This connection between process and product is one of the recurring themes of the book.

The introduction of metallic glasses is indicative of advancements in thin film coatings. Owing to their amorphous structure, metallic glasses exhibit improved properties that traditional crystalline materials do not offer. Properties such as corrosion resistance, hardness, and wear behaviour among others are improved due to the lack of long-range atomic order. These metallic glass coatings are obtained by laser cladding or thermal spraying. Their inclusion in this volume shows the importance of continually looking for new material systems instead of just depending on established coatings.

Defects remain a pertinent issue in thin film coatings during deposition. Defects are unavoidable; however, controlling them to a larger extent can be achieved. Be it point defects, line defects, or the bulk ones, they control the performance of the thin film coatings. Hence, characterization techniques play an important role in understanding this issue in detail. The chapters dealing with the defects in coatings provide a comprehensive idea to the readers on the subject of stresses, effect thermal stresses, and deposition parameters, among others, which are important attributes that control defects.

Another strand comes from observing nature. Surfaces in the natural world display remarkable behaviours, such as water repelling on lotus leaves and strong attraction on other biological surfaces. These natural patterns of hydrophobicity and hydrophilicity have guided the design of thin films with engineered wetting properties. By changing roughness and surface energy, coatings can be tailored for self-cleaning functions, biomedical applications, or specialized industrial processes.

The shift from natural inspiration to technological application is an important part of current thin film research. Its coverage here shows how closely science often follows observation.

In recent times, spray pyrolysis has become particularly relevant in the field of the thin film coatings. Interestingly, this method was developed for making powders but tweaked into depositing thin films. One of the attractive features of this process is its simplicity; it does not require a vacuum set-up. This process has found its application in the field of energy applications, catalysis, or sensing. Introducing this process in this volume gives readers a new perspective on the developments in the area of thin film coatings. The future of thin film coatings lies in smartness. Smart coatings that can detect defects, repair themselves, and respond to changes are creating paradigm shifts in this field. Their applications in oil and gas sectors are suggestive of the future where coatings do not only simply protect but plays an important role in monitoring the system or process.

Equally important are detailed studies of specific coating systems. Titanium nitride and aluminium chromium nitride are established materials; yet their properties can be enhanced dramatically by small changes. The introduction of interlayers or silicon additions produces films with higher hardness, better adhesion, and improved wear resistance. In biomedical applications, surface energy modification can even reduce bacterial adhesion. These focused investigations show how incremental but careful modifications in microstructure or composition can transform industrial performance.

The closing discussions point towards the integration of tribology with modern data analysis, AI, and machine learning. Wear has long been measured in terms of simple averages, but new approaches use wavelet analysis, entropy measures, and machine learning to capture the complexity of wear regimes. Running-in, steady-state, and unstable wear behaviours can now be distinguished in fine detail. This represents a merging of materials science with data science, opening opportunities for predictive models and real-time monitoring.

Taken together, the chapters in this book follow the arc of thin film coatings from their origins to their present directions. The contributions are linked by common themes: the tension between bulk and surface, the constant push to refine deposition, the acceptance of defects as challenges to be managed, the inspiration drawn from natural systems, the balance between simplicity and sophistication, and the growing reliance on data to understand complex behaviours.

This book brings together perspectives from across the field to provide a wide view of where thin film coatings stand today. It does not treat each topic as an isolated theme but rather as part of an ongoing story where older techniques coexist with new materials and where practical problems drive scientific innovation. Thin films remain only nanometres thick, but the science surrounding them continues to broaden, showing once again that some of the most important changes in engineering begin at the surface.

Acknowledgements

This book has been a journey made possible by the support, generosity, and encouragement of many people. I would first like to express my deepest gratitude to the contributing authors, whose expertise and dedication brought depth and perspective to each chapter. Their willingness to share knowledge and experience has been invaluable in shaping this volume.

I am also indebted to my colleagues and mentors, who offered guidance, thoughtful discussions, and constructive feedback at different stages of this project. Their insights not only strengthened the content but also inspired me to view thin film coatings through new and broader lenses. My sincere thanks also go to the reviewers for their patience, careful reading, and commitment to clarity and quality, which helped transform this work into a cohesive whole.

I would like to extend my heartfelt thanks to my co-editors – **Dr. Sunil Kumar, Dr. Saikat Ranjan Maity, and Dr. Magdalena Łępicka** – for their collaborative spirit, intellectual contributions, and unwavering support throughout this endeavour. Their expertise and commitment have been central to shaping the quality and coherence of this book.

I am also grateful to **Prof. Kapil Gupta**, Series Editor of *Advanced Materials Processing and Manufacturing*, for his encouragement and for providing the platform that made this book possible. His vision for advancing research in materials science has been an inspiration to all of us involved in this work. On a personal note, I am profoundly grateful to my family for their constant encouragement and understanding.

Above all, I dedicate this book to my late father, Shri. Kameswar Patnaik, whose values, love, and inspiration continue to guide me every day.

Lokeswar Patnaik, PhD

1 An Overview of Thin-Film Coating Technologies

Abhik Bose, Himanshu Batham, and Tushar Banerjee

1.1 INTRODUCTION

Thin film deposition processes involve techniques for creating surface layers, with thicknesses ranging from several tens of micrometres to nanometres. Current trends show that a film's thickness can be reduced to a nano regime layer. A suitable material's surface, known as the 'substrate,' is required for the deposition of such films. The coating process essentially includes depositing a thin film of the material onto a base, which can be done using either liquid-phase or vapour-phase methods [1]. Various coating techniques can be tailored to specific production needs, such as controlling layer thickness, surface roughness, deposition rate, and the size of the coated product [2]. Film width, patterning capability, and coating speed are some of the factors that affect these characteristics. The coating process is classified depending upon the volatility of the precursor that can take place in low vacuum or ambient pressure.

Solution coating involves applying a solution directly onto the substrate, dissolving the matrix material and binder in it. A dry film is left over after the solvent evaporates. On the other hand, thin, conductive [3], semi-conductive [4], and dielectric layers with particular morphologies typically develop by vacuum deposition techniques, such as CVD (via chemical route) [5], PVD (via physical deposition route) [6], atomic-layer deposition (ALD) [7], plasma [8], flame hydrolysis deposition (FHD) [9, 10], and sputtering [11]. These thin films are crucial for heat or electricity transfer and are generally produced under high-vacuum conditions. Figure 1.1 presents a classification of the widely used vapour and liquid phase deposition processes. The main objectives of coating technologies include (i) applying conductive or insulating layers to a surface, (ii) creating localized conductive channels to connect devices, (iii) constructing complex layer structures through multiple passes. Over the years, various technologies for coating deposition and thin film preparation have been developed [12–14]. These methods are especially advantageous for solid-state device fabrication because they use significantly less material compared to other techniques that utilize larger quantities of bulk materials, thereby reducing processing costs. Coating deposition offers several benefits, such as surface uniformity, ease of fabrication, lower production costs, and unique opto- and mechanical properties.

DOI: 10.1201/9781003750659-1

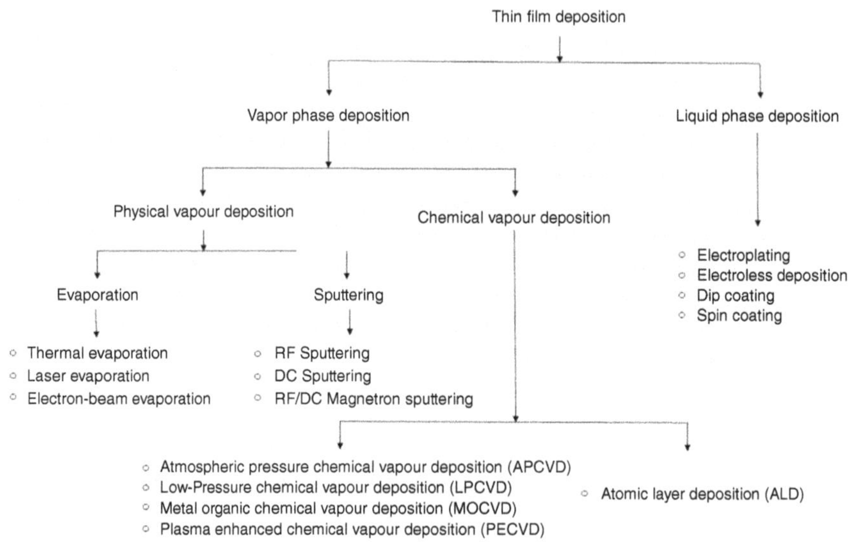

FIGURE 1.1 Flow chart illustrating the most widely used deposition techniques.

1.2 DEPOSITION PROCESSES BASED ON PHYSICAL PHENOMENON

Some key and cutting-edge technologies based on vacuum-assisted physically material synthesis techniques are discussed. PVD is a technique of knocking out atoms/ions from a source and simultaneously depositing them on a base substrate to form a coating film involving or without involving any chemical reactions. All PVD techniques are done in a vacuum chamber, which can control the atmospheric conditions inside. Some of the widely used methods for PVD are evaporation and sputtering.

1.2.1 Evaporation

Faraday invented the novel evaporation technique for thin film deposition in the 1850s, now called thermal evaporation [15]. This is one of the simplest methods of coating a substrate which usually uses a source to make thin films by thawing within an idle surrounding. The sample is heated within the conventional vacuum coating machine to its vapour pressure. The substance moves from the sample holder to the substrate holder as it evaporates, covering the substrate's surface.

Thermal evaporation can be used to produce slim films of both metallic and non-metallic materials, mainly aluminium, gold, silver, and indium. This technique is often employed in industrial manufacturing to create electrical contacts using materials like silver, copper, or aluminium. The typical set-up for a thermal evaporation apparatus is presented in Figure 1.2(a). The specimen is placed in a red-hot boat, and the process uses resistive heating to evaporate the sample's atoms, atom clusters, or molecules. The substance is either directly or indirectly vapourized in the sample container using an electric arc or a heating filament made of carbon or molybdenum.

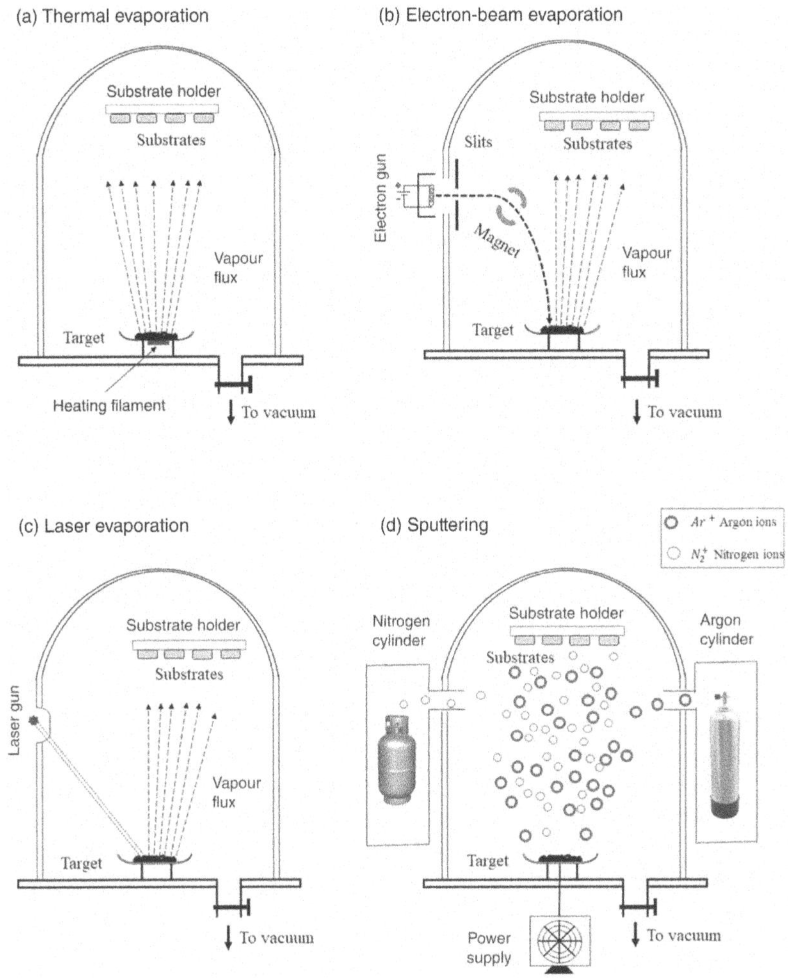

FIGURE 1.2 Different kinds of physical vapour deposition processes: (a) thermal evaporation, (b) electron-beam evaporation, (c) laser evaporation, and (d) sputtering.

The sample is positioned at the apex along with a clamp, and all components are enclosed in a vacuum chamber or using a glass vessel. Thermal evaporation is particularly useful for depositing metal coatings as contact layers in integrated circuits, such as organic LEDs, thin-film transistors, and solar cells. Typical filament currents range from 200 amps to 300 amps, and deposition rates are between 1 Å per second and 20 Å per second. These evaporation techniques are effective for temperature range up to 1800 °C. Although there are numerous advantages of thermal evaporation methods such as ease of deposition, less expensive, purity of films, and so on, one of the significant backlogs of such process is the wastage of material during the deposition [16]. Owing to high-temperature operations, the process is also not suitable for deposition of coatings on substrates with low melting points.

Figure 1.2(b) depicts the electron-beam evaporation technique, also known as the second evaporation technique. This technology was initially discovered by Pirani [17] incidentally and subsequently advanced by Ruhle [18]. This technique involves directing a concentrated stream of electrons towards the sample material in the container, resulting in the generation of a vapour flow. The trajectory of the electron beam is magnetically manipulated to precisely aim at the sample. The emission voltages commonly used for this approach fall within the range of 8 to 10 kV, while the deposition rates might vary from 0.2 to 100 Å/sec. An advantage of Electron Beam Evaporation is its capability to interchange within a wide range of targets into the trajectory of the e-beam without compromising the vacuum. This technology is widely employed in optical applications like solar panels, laser optics, architectural glass, eyeglasses etc. The device can deposit a diverse array of materials, which is essential for achieving important mechanical, electrical, and optical characteristics. In comparison to other (PVD) methods, electron beam evaporation demonstrates superior efficiency in resource usage, resulting in reduced production costs.

This challenge is addressed by the third method, known as laser evaporation, depicted in Figure 1.2(c). Instead of utilizing an electron gun, this method uses a laser gun to evaporate the sample material. Usually using a laser light source, it's referred to as laser pulse deposition. A tiny amount of the sample evaporates in a thermally energetic burst when this energy source is absorbed by the target's surface. This method's main benefit is that the target material's atomic ratios are preserved both before and after evaporation, as the evaporated sample is placed on the substrate with little degradation. When depositing materials with complex stoichiometry is vital, this method is especially advised. Using a vibrating crystal, the surface profile of a thin layer produced by evaporation can be continually monitored throughout the experiment. The material being deposited onto the vibrating crystal causes a change in its frequency, which is the basis for this reading. As more material is deposited, the resonance frequency decreases.

1.2.2 Sputtering

The principle of sputtering was initially captured by William Robert Grove in the mid-19th century [19]. However, it remained mostly a scientific inquisitiveness until the 1940s, till diode sputtering began to be used as a commercial deposition process. Sputtering involves using energetic ions to knock out molecules or atoms off a source surface, which then deposit on the surface to form a thin layer. For a good sputtered film, two main criteria are essential:

- The energy of the incident ions must be higher than the binding energy of the atoms being sputtered; otherwise, the ions will simply be reflected or absorbed by the surface.
- The ejected atoms need to move freely towards the substrate with minimal disturbance to their motion.

Sputtering is essentially a high-vacuum process, requiring minimum pressures to produce high ion energies and minimize collisions between atoms and gas molecules, thus increasing the mean free path. Pressures below 10–2 mbar or 1 Pa are ideal. Higher pressures lead to more gas collisions and lower deposition rates. Sputtering yield can be optimized by tuning the charge density of the incident ions and the weight of the colliding atoms, with silver having the highest yield among metals. Figure 1.2(d) illustrates a basic DC sputtering chamber. The key components include a vacuum chamber, an electric supply, and a carrier gas. Inert gases like argon, krypton, and xenon create the plasma, with argon being the most commonly used because it does not chemically or physically react with the target material and enhances the coating quality. The super-ionized gas is generated by heating the chamber gas. Sputtering offers several advantages:

- It can coat large-area targets, providing uniform thickness across the wafer.
- Thickness can be tuned by controlling and modifying the coating time and other parameters.
- Materials with very high melting points can be sputtered.
- The cohesive bond between the sputtered thin film and the substrate is strong, ensuring good quality.
- Sputtering can be performed in both vertical and horizontal orientations.

However, there are some drawbacks to sputtering. The ions generated have notable kinetic energy (~0.5–1 keV), much higher than the energy of the atoms sputtered in the process (1–10 eV). This means that most of the incoming energy (>90%) is wasted as heat, making sputtering less energy-efficient. Additionally, a well-organized cooling system is crucial to prevent damage to the target material due to excessive heat caused by the higher sputtering rate during deposition processes. Further, there is an inherent 'line-of-sight' issue which refers to deposition of coating material on surface only visible to the sputtered atoms or ions.

1.2.3 Magnetron Sputtering

Sputtering machine set-ups differ based on the films to be used and the work environment in which they will be used. The mechanical characteristics and shape of the film can be significantly influenced by the power supplies utilized. Furthermore, sputtering efficiency is determined by the flow of electrons. The sputtering yield in diode magnetron sputtering systems is usually low because of the argon gas's inefficient ionization. By using three main electrodes with independent power sources – one for cathode heating, one for plasma voltage, and one for the high voltage supplied to the sputtering target – a triode sputtering system, on the other hand, increases sputtering yield. By employing magnets to increase electron collisions with sputtering gas atoms, modern developments have improved electron participation in ionization events. Both the electrons' ionization capability and route length significantly increase as a result. Magnetron sputtering is the name for this sophisticated form of sputtering. One advantage of using magnets is that it requires less voltage to strike the plasma. Deposition rates are increased and are better regulated. The charring of the base caused by electron collision is decreased. A schematic illustration of planar magnetron sputtering is presented in Figure 1.3. An electron trap is created in a

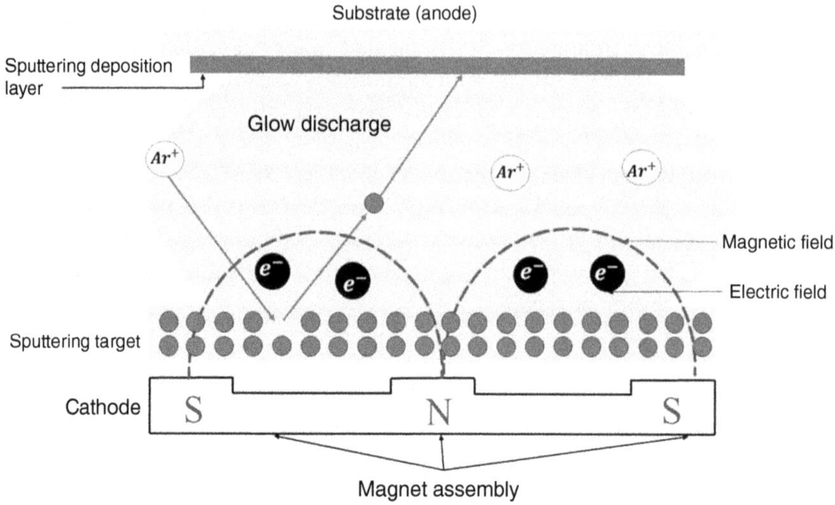

FIGURE 1.3 Schematic of planar magnetron sputtering.

magnetron-enhanced diode by immersing the cathode in a magnetic field. The electron drift current E × B (electric field strength × magnetic flux density) is contained within a closed loop on the surface by this electron trap. An unending 'racetrack' contains the primary and secondary electrons.

Both the quantity and the efficiency of ionizing collisions are effectively improved by this racetrack. Magnetron mode is the name given to this cathode operation technique. Magnetrons come in a variety of designs, including planar, ring or gun, cylindrical, and more, all of which operate on the same fundamental principle.

DC magnetron and RF magnetron sputtering are the two primary categories of sputtering. Magnets are positioned within the negative-biased source material to confine the atoms over the cathode in both techniques. Direct current is used in DC magnetron sputtering to provide electricity. Vacuum levels are maintained of the order of 10^{-5}–10^{-7} Torr. It is usually applied to materials with good electrical conductivity, which lowers expenses and boosts efficiency. This technique offers high deposition rates for particular metallic targets and is simple to use, even for large substrates. In DC sputtering, the target is reached by the quickly travelling positively charged ions from the chamber gas, and the expelled atoms collect on the substrate surface for deposition process. An AC current is used in RF magnetron sputtering, where a radio frequency source enables a strong electric supply at a steady frequency of 10–15 MHz. This method works well with dielectric targets and may be applied to both conductive and non-conductive materials. To save processing costs, RF sputtering is usually employed for smaller substrates because its deposition rate is slower than that of DC sputtering. The target functions as a cathode during the negative half-cycle, drawing polarized sputtering gas atoms to eliminate its atoms. To accelerate source atoms and gas ions towards the substrate for coating deposition, the target turns into an anode during the positive half-cycle, reversing the polarization.

1.2.4 Balanced and Unbalanced Magnetron Sputtering

Depending on the magnetron arrangement, the process can be classified as either balanced or unbalanced magnetron sputtering [20]. When the inner and outer magnets have the same strengths, the magnetic field is deemed to be in a balanced state. The magnetic field lines are closed during balanced magnetron sputtering, which stops electrons from leaving the magnetic field. As a result, plasma forms around the cathode, and argon is ionized, producing a powerful plasma within near proximity of the target surface. The main purpose of developing this balanced magnetron was to prevent the bombardment of energetic particles on the growing layer, especially in microelectronic applications thereby reducing stresses in the film. Nevertheless, balanced magnetron sputtering is subject to two notable constraints: it yields ion current densities below 1 mA/cm^2 and confines the plasma primarily near the cathode, hence limiting its effectiveness in activating reactive gases near the substrate for reactive sputter deposition. On the other hand, unbalanced magnetrons exhibit a magnetic field that is not evenly distributed, resulting in certain field lines not being connected across the central and outer poles. Two types of magnet configurations can be utilized in unbalanced magnetron sputtering: (i) a weaker central magnet with stronger ones at the periphery, and (ii) a stronger middle magnet with weaker ones at the edges. In the first case, certain magnetic field lines extend towards the substrate, enabling secondary electrons to trace their path. The spreading of the plasma towards the substrate leads to greater ion currents. In addition, the ions that reach the substrate have low energy, making them well-suited for reconstructing the coating and achieving the best deposit characteristics. Since the magnetic field lines in the first configuration are not restricted to the core area, electrons can escape from the plasma and move in the direction of the substrate. The second configuration lowers the plasma ion charge energy in the substrate by extending the lines of the unclosed magnetic field towards the chamber walls.

1.2.5 Closed-Field and Mirror-Field Unbalanced Magnetron Sputtering

Although unbalanced magnetron sputtering offers benefits, it is still difficult to get even coatings on substrates with intricate geometries like cutting tools at satisfactory rates with only one source. Moreover, a single cathode system is insufficient for sizeable tools such as broaches, leading to the formation of stratified layers despite the rotation of the substrate. To overcome these constraints, a system of multiple magnetrons has been devised. The magnetic arrays in neighbouring or facing magnetrons can be arranged with either the same or opposite magnetic polarity in this arrangement [21]. When the magnetic polarities facing each other are the same, the arrangement is said to as 'mirrored.' The field lines are directed towards the walls of the chamber. Secondary electrons that travel along these paths become depleted, leading to a reduced plasma density in the vicinity of the substrate. On the other hand, when the magnetic polarities facing each other are opposite, the arrangement is called a 'closed-field.' In this configuration, the field lines establish connections between the magnetrons, reducing losses to the walls of the chamber and positioning the substrate within an area of high-density plasma. The closed-field design creates a magnetic trap that efficiently confines electrons between the cathodes, leading to high concentrations of ion current and significant bombardment of the expanding film by ions. Closed-field unbalanced magnetron sputtering enables the generation of extremely high levels of substrate bias current densities, which are crucial for producing totally compact films with relatively minimal residual stress during the reactive deposition of hard compounds. This method is widely acknowledged as a high-speed method for depositing metallic coatings.

1.2.6 Pulsed Direct Current (DC) Magnetron Sputtering

Another form of PVD coating method known as pulsed direct current (DC) magnetron sputtering is commonly used to manufacture metallic coatings with precise mechanical, electrical, and optical properties [22–24]. The primary benefit of pulsed DC sputtering over conventional DC sputtering is arc suppression. Unwanted arcs, which could otherwise degrade the thin film quality, are greatly reduced or even eliminated by this procedure. An intense negative pulse is applied to the target to start sputtering. This causes collisions between the coating material and the target material, resulting in the creation of an ionized gas. A low-voltage, brief cycle reversal that 'scrubs' the target surface in advance of the subsequent sputtering pulse comes next. A negative bias is typically used in the pulsed DC sputtering process, and the pulsing frequencies range from 10 kHz to 350 kHz [25]. This band of frequencies aids in preventing arc development at the target. However, since mid-frequency pulses (100–350 kHz) may provide better control over the ion current, there is growing interest in the advantages of applying these pulses [26]. Two types of pulsed DC magnetron sputtering are commonly identified: unipolar and bipolar pulsed processes. In unipolar pulsed sputtering, the target surface is cleaned, and the build-up of a dielectric layer is stopped by switching off the pulse for a certain duration on the power waveform. To further reduce the effects of dielectric build-up, bipolar pulsed sputtering uses two 180-degree out-of-phase pulses, with each magnetron alternating between cathode and anode roles. Several variables, including pulse length, frequency, power, and pressure, affect how long plasma forms and stabilizes. The plasma accumulation phase predominates with shorter pulses and higher frequencies, and pulsed magnetron sputtering runs in voltage mode. On the other hand, the stationary plasma phase is more noticeable, and the pulsed magnetron runs in current mode at longer pulse lengths and lower frequencies. In the mode of constant current, the ions are rapidly accelerated towards the target by a high voltage, which maintains the deposition rate and restores the current.

1.2.7 High-Power Impulse (HiPIMS) Magnetron Sputtering

In recent decades, various magnetron sputtering technologies have been developed to produce dense ionized fluxes of target material while enhancing control over ion bombardment [27, 28]. Among these advancements is the technology referred to as high-power impulse magnetron sputtering (HiPIMS) or high-power pulsed magnetron sputtering (HPPMS), which was first described by Kouznetsov et al. (1999) [29] as a way to enhance magnetron sputtering utilizing pulsed-power technology [30].

As per Anders' [31] explanation, HiPIMS occurs when the peak power is roughly two times higher than the time-averaged output. This method has been industrialized by a number of major coating companies under various acronyms, which is a positive step towards its integration into standard thin-film processes. The main feature of HiPIMS is the integration of periodic plasma discharges with conventional magnetrons for sputtering, resulting in a highly ionized plasma including sizeable amounts of ionized sputtered particles [32]. Moreover, HiPIMS has been shown to reduce deposition temperatures, enhance film adherence, and enable the coating of uniform layers on substrates with intricate shapes [33–42]. It is worthwhile to create novel ionized sputtering techniques like HiPIMS because of these advantages. HiPIMS may apply several hundred kW, resulting in larger ion charge energy and increased ionization of the target material, compared to a standard sputtering power supply that ranges from 0.5 kW to 10 kW. The following are variations in the deposition procedure:

- Target material ionization, with 70–80% of the material ionized.
- The deposition rate can be slowed down by redirecting metallic ions from the target back to it.
- More energy and mobility of the ions when they reach the substrate result in thicker, more compact films.

The low deposition rate and considerable residual stress created in the films using HiPIMS are two significant disadvantages. The number of magnetrons in the chamber is being increased to overcome this problem. Recently, Nano4Energy in Spain created a new HiPIMS variation that improves the deposition rate by allowing deposition during the HiPIMS pulse's off-time.

1.3 DEPOSITION PROCESSES BASED ON CHEMICAL PHENOMENON

In this section, various notable and sophisticated processes based on chemical deposition process technologies will be explored. In general, chemical deposition techniques require the transformation of a gaseous precursor into a solid layer on the surface of a solid object. The precursor gases are often mixed with carrier gases and then injected into the reaction chamber at temperatures close to the surrounding environment. These gases undergo chemical reactions or decomposition when they come into contact with a heated surface, resulting in the formation of solid deposits on the surface.

1.3.1 Chemical Vapour Deposition

Chemical vapour deposition (CVD) is a collection of methods in which a solid material is deposited onto the surface of a sample by a chemical route [43]. CVD is based on the principle of decomposing pure gases in a controlled chamber. This process leads to a sequence of chemical and physical processes, ultimately ending in the formation of a solid layer on the substrates. This process is highly adaptable and

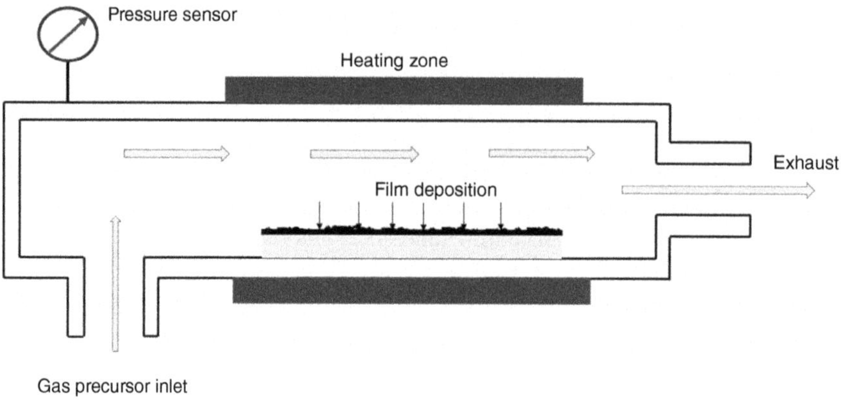

FIGURE 1.4 Schematic for chemical vapour deposition.

effective, allowing for the production of defect-free coatings with a consistent size and controlled porosity, even on complex or curved surfaces. CVD is very suitable for the industrial production of thin-film fibres and the coating of uniform components. By manipulating several experimental variables, including the substrate material, temperature, gas composition, and pressure, we can create thin layers with a vivid variety of desirable qualities suitable for varied applications. To achieve efficient CVD processing, the precursors must possess both volatility and stability, allowing them to be easily delivered to the reactor. Chemical vapour deposition is a technique where gaseous precursor molecules react chemically to form solid thin films by depositing coatings onto a suitable substrate. CVD is capable of depositing both elemental and compound coatings from compound gaseous precursors. In the subsequent sections, we will examine some notable categories of CVD techniques. Figure 1.4 shows the schematic of deposition by the CVD route.

The CVD process can be clearly described in the following steps: (i) a controlled flow of 'Certain reagents and inert gases' is introduced into the reaction chamber, (ii) the gases disperse and spread across the surface, (iii) high temperature of the substrate triggers chemical reaction leading to the formation of the reaction product and its deposition in the form of a film, (iv) the by-products of the reaction are removed from the chamber through desorption and evacuation. Choosing suitable precursors is essential for attaining the intended coating. In classical CVD systems, metal hydrides and halides have traditionally been employed as precursors. Recently, a diverse range of metal-organic compounds, such as metal alkyls, metal alkoxides, metal diketones, metal amidinates, and metal carbonyls, have been used as chemical precursors.

1.3.2 Atmospheric Pressure Chemical Vapour Deposition (APCVD)

Atmospheric pressure chemical vapour deposition, abbreviated as APCVD, was the initial process to be developed in this particular discipline. The process is simple, makes use of affordable equipment, and enables fast deposition rates. Nevertheless, because it is prone to gas-phase interactions, it necessitates high gas flow rates. This technique is efficient in generating relatively thick layers on different substrates or wafers, with thin films typically measuring several micrometres in thickness. APCVD is useful for refining the surfaces of various materials, such as tools and turbine blades, hence improving their longevity and effectiveness. An advantage of this CVD process is its ability to operate without the need for a vacuum, resulting in a relatively lower cost. However, APCVD is not commonly employed because of the elevated danger of oxidation caused by the increased gas density and prolonged residency durations. Moreover, it fails to offer consistent coverage on uneven surfaces featuring steps, and there is a potential for contamination by particles.

1.3.3 Low-Pressure Chemical Vapour Deposition (LPCVD)

Low-pressure chemical vapour deposition (LPCVD) reactors offer higher film quality in comparison to APCVD reactors, resulting in improved uniformity, step coverage, and less particulate contamination. During the LPCVD process, pressure is carefully controlled to be much lower than the pressure of the surrounding atmosphere, typically around 100 Pascal (Pa). This leads to a decrease in the pace at which particles spread out by a factor of 1,000, which in turn decreases the rate at which mass is transported and enables reactants to come closer to the surface. Consequently, LPCVD coatings demonstrate exceptional uniformity, evenness, purity, and ability to cover steps. The application of low pressure reduces unwanted gas-phase interactions and improves the evenness of the coating on the substrate. Lowering the pressure enhances the movement of precursor materials through the gas phase, hence increasing the pace at which gaseous reactants are transported and surpassing the rate at which surface reactions occur. Nevertheless, LPCVD necessitates intricate apparatus, functions at elevated temperatures, and exhibits a reduced deposition rate.

1.3.4 Metal Organic Chemical Vapour Deposition (MOCVD)

Metal organic chemical vapour deposition, abbreviated as MOCVD, is a method used to coat a thin layer of metal-organic compounds onto a substrate by introducing a vapourized precursor into a reaction chamber. MOCVD is a specific variant of LPCVD that takes place under moderate pressures, which are close to vacuum conditions. It is commonly used for the development of epitaxial layers. This procedure entails the utilization of organometallic chemicals that readily undergo decomposition at moderate temperatures. Manasevit (1968, 1972) [44, 45], Manasevit and Simpson (1969) [46] played a crucial role in the development of the MOCVD method. They effectively utilized metal-organic compounds to produce thin films of different semiconducting and crystalline materials. MOCVD utilizes metal-organic chemicals as molecular precursors to deposit coatings, enabling the production of diverse thin films for advanced scientific and technological purposes. One notable benefit of MOCVD is its ability to produce thin films of metals and metallic compounds, such as metal nitrides and metal oxides, by utilizing precursors with high vapour pressure at lower reaction temperatures, distinguishing it from other CVD techniques.

1.3.5 Plasma-Assisted Chemical Vapour Deposition (PACVD)

Plasma-assisted chemical vapour deposition (PACVD) is a highly common form of CVD. PACVD, instead of depending entirely on elevated temperatures to facilitate chemical reactions, employs plasma to supply a portion of the required energy. The ionization of gas in the plasma facilitates the reactivity that is usually attained at elevated temperatures, hence enabling deposition at significantly lower temperatures. The temperature range used in this process is lower than that of APCVD and LPCVD, making it appropriate for depositing coatings on surfaces with low melting points. PACVD offers significant benefits in the silicon sector for the application of amorphous silicon films and coatings on silicon wafers. This includes the deposition of compounds such as silica, silicon nitride, and silicon oxynitride. PACVD exhibits a high level of reactivity, enabling it to effectively treat materials at reduced temperatures and pressures. PACVD enables the utilization of substrates that have little tolerance for elevated temperatures (ranging from 373 K to 773 K), such as polymeric materials. Nevertheless, thin films fabricated using PACVD typically exhibit lower quality compared to those generated by LPCVD. PACVD-produced films frequently exhibit non-stoichiometric composition, with the film incorporating chemical by-products such as hydrogen, oxygen, and nitrogen. Excessive contamination might result in the occurrence of film cracking and delamination when subjected to repeated heat cycles.

1.3.6 Atomic Layer Deposition

Atomic layer deposition (ALD), developed by Suntola (1989), employs consecutive and self-saturating chemical surface reactions to create thin films [47]. This is achieved by introducing gaseous reactants, also known as precursors, into the reaction chamber. Originally known as atomic layer epitaxy, ALD is distinguished by the intermittent delivery of precursors in a specific order, with intervals of inert gas purges to avoid reactions in the gas phase. ALD involves the sequential exposure of the substrate surface to at least two highly reactive precursors [48]. Each precursor contains specific components of the desired material. These precursors control the surface saturation of the sample monolayer. The majority of commercial ALD systems employ an inert carrier gas, such as nitrogen (N_2), at an appropriate level of vacuum, often around 1 millitorr. The precursors are injected in small increments into the carrier gas stream, and these increments are then followed by brief purging intervals to eliminate any excess reactants. The sequence of each ALD cycle normally consists of a first precursor pulse, followed by a purge, then a second precursor pulse, and finally another purge. In each cycle, a specific quantity of material is deposited as a single layer, which is approximately 1 Å in thickness. The interaction amongst the precursor molecules and the surface occurs self-limitingly, ensuring that the reaction stops once all available reactive sites on the substrate are occupied. The design of each ALD cycle considers the characteristics of the interaction between the precursor and the surface. The process can be iterated numerous times to accumulate the necessary quantity of thin film layers.

1.3.7 ALD Versus CVD Deposition Technique

ALD is a type of deposition technique that belongs to the category of CVD. However, it distinguishes itself from CVD by exposing the substrate to distinct precursors in a sequential manner rather than all at once. CVD involves the heating and continual introduction of volatile chemicals into a deposition chamber. These compounds chemically react on a heated substrate, resulting in the formation of a film composed of the desired material. Single-source precursors (SSPs) can incorporate many components into a film using only one precursor molecule. However, this process might lead to the presence of carbon impurities due to the burning of organic ligands. Alternatively, distinct reactive gases known as co-reagents can be utilized to facilitate the appropriate chemical reactions in the film formation process. Typical co-reagents for the formation of oxides are oxygen, water, or ozone. It is extensively employed in the microelectronics sector to produce interconnects and semiconductor materials. However, an inherent disadvantage of CVD occurs when dealing with intricate substrates. Overgrown high-aspect-ratio structures can occur during deposition because of the increased likelihood of reaction near the top of the feature caused by thermally unstable precursors. As a result, there is an uneven distribution of material, causing the lower parts of the features to receive less coverage. Therefore, CVD is not suitable for applications that necessitate uninterrupted, uniform coatings over complex structures, such as in photovoltaics, memory storage, or hydrophobic materials. CVD is preferred for simpler substrates compared to line-of-sight techniques like sputtering despite certain constraints. Sputtering typically fails to uniformly cover the surfaces of nanostructures. CVD provides enhanced precision in regulating film roughness and conformality in comparison to sputtering, which is necessary for achieving films of exceptional quality. ALD, while sharing similarities with CVD, is especially adept at covering intricate surfaces. The main distinction lies in the fact that ALD utilizes volatile precursors that undergo surface reactions in a self-limiting manner. The reaction ceases when the entire surface of the substrate is coated, as there are no additional catalytic sites available for the precursor to interact with, resulting in a self-terminating or self-limiting process. ALD, albeit slower than CVD, provides the benefit of conformal coverage on surfaces with large areas or aspect ratios due to its self-limiting nature. This rigorous procedure guarantees consistent coatings, even on extremely complex surfaces. Figure 1.5 illustrates the various types of coating achieved on high aspect ratio feature surfaces through the PVD, CVD, and ALD processes

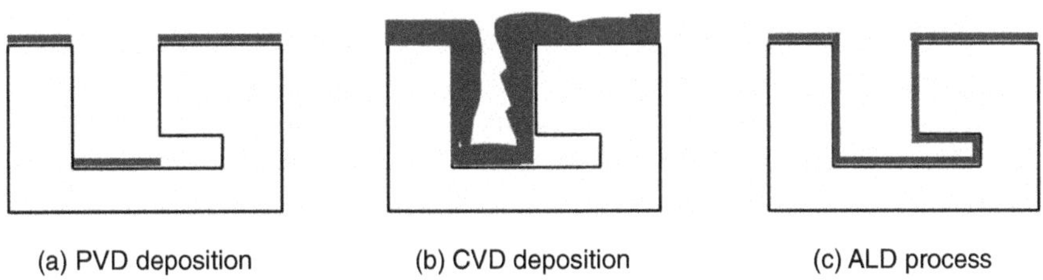

FIGURE 1.5 Illustration of edge coverage for PVD, CVD, and ALD.

1.4 TECHNOLOGIES DEPENDING ON LIQUID PHASE DEPOSITION

1.4.1 Electrodeposition

Electrodeposition, commonly known as electroplating, is the process of applying a substance onto an electrode by using electrolysis. This technique is widely employed in numerous industries to increase the attributes of materials, such as increasing corrosion resistance, enhancing aesthetic appeal, or imparting certain surface features [49]. By applying an electric current to an electrolyte solution, metal ions are reduced and subsequently deposited onto a conductive surface. This procedure enhances the visual appeal of materials and also enhances their durability and performance. Faraday's rules govern the method when a material is applied to an electrode through the process of electrolysis. Electrodeposition or electroplating is founded upon this principle. Chemical reactions are induced by the flow of an electric current via an electrolyte in this process. The magnitude of these chemical alterations is closely correlated with the intensity of the electric current. The number of different species that accumulate or dissolve on the electrodes is directly proportional to their corresponding weights and the electric current applied. Electroplating commonly involves the use of three electrodes (Figure 1.6(a)). The working electrode (WE) is constructed using the material that will be used for plating. The reference electrode (RE) is an electrode that senses voltage and maintains a consistent electrochemical potential. The counter electrode (CE) serves as the pathway for the electric current, thereby ensuring the completion of the electric circuit. The current passing through the RE cannot be regulated independently, as it is dependent on the desired process at the WE. Therefore, the employment of the CE is necessary. The electrodes are connected to a mechanism that ensures a constant potential during deposition and controls the electrodeposition rate [49]. A potentiometer is commonly employed for this objective.

1.4.2 Electroless Deposition

The history of the process known as electroless deposition dates back several centuries. Silver and gold were electroless deposited by ancient civilizations, albeit they were unaware that they were doing so. Significant scientific development was accomplished in the second half of the 20th century, despite gradual advances over the ages. As shown in Figure 1.6(b), the aqueous electroless bath essentially consists of metal salts (source of metal ions), reducing agent, complexants, stabilizers, and pH buffers [50–56]. Before immersion in the bath, the substrate needs a series of pre-processing steps consisting of polishing, degreasing, rinsing, and activation. The reduction of the metal salt is triggered by the autocatalytic substrate immersed in the bath. The electroless process can be used to deposit metals, compounds, poly-alloys, multi-layers, composites, and graded coatings on both conductive and nonconductive substrates.

1.4.3 Spin Coating

A simple chemical deposition method called spin coating can be used to cover a variety of materials, most commonly polymers, uniformly. To produce uniform coatings

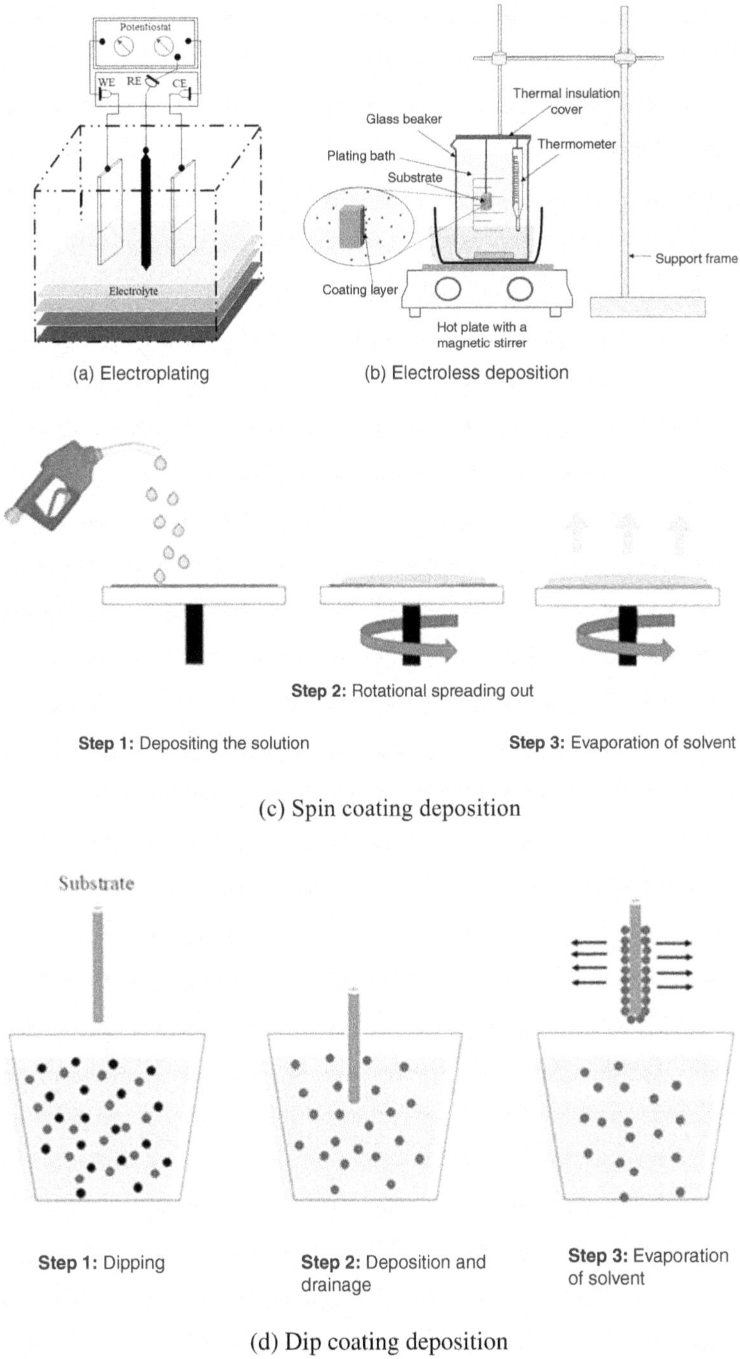

FIGURE 1.6 Illustration depicting the functioning of (a) electroplating, (b) electroless deposition, (c) spin coating deposition, and (d) dip coating deposition.

with the required thickness, spin coating is used. But this approach is ineffective because it wastes more than 90% of the resources. As a result, as the film's coverage area grows, so does the cost of supplies [57]. Spin coating has been used for applying thin films for many years. With this method, the substrate is rotated at a regulated high speed after a small quantity of coating material is placed in its centre. The substrate revolves about an axis that is perpendicular to the area being coated for the spin coating process to occur. Consequently, as the material used for coating is spread towards and subsequently moves away from the substrate's edge, a thin coating layer forms on the surface. The coating's qualities (drying rate, viscosity, surface tension, percentage of particles, etc.) and the spin processing parameters (duration, rotation speed etc.) define the final film thickness and other features. Figure 1.6(c) illustrates the many steps of the spin coating process. One major constraint on the spin coating process is the substrate's dimensions. The difficulty of thinning the film grows with substrate size, making high-speed spinning more challenging. Regarding material efficiency, spin coating shows a considerable restriction. Roughly 95–98% of the material is usually squandered during the entire process, leaving approximately 2–5% of the material to be put to the substrate.

1.4.4 Dip Coating

The fourth technique, which is both basic and cost-effective, is dip coating [58, 59]. This process involves four phases as shown in Figure 1.6(d). Initially, a substrate is immersed vertically at a consistent velocity into a container filled with the coating solution. To obtain a symmetrical and uniform deposition, the substrate is traditionally dissolved vertically. During the second stage, the remaining substrate is immersed in the solution for a specified duration. The deposition technique commences on the substrate during the third step, as soon as it is withdrawn following the completion of the immersion time. The key elements that dictate the coatings' thickness are the rate at which the sample is inserted and removed, as well as the length of the immersion. The last stage is to drain the extra liquid and evaporate the solvent from the substrate's surface, which forms the thin layer of the coating material. Submerging the substrate many times allows for controlled multi-layer deposition. Multi-layer structures with varying ingredients in each layer can be created using multivessel dip coaters and different dipping solutions. The efficacy of the dip coating process can be directly influenced by various parameters, including pH and solution concentration, which can be altered to modify its viscosity.

1.5 CONCLUSIONS

Thin film coating technologies aim to bring in necessary surface characteristics to the substate through deposition of layers which may range from a few nanometres to tens of micrometres. Owing to such small thicknesses, the original geometry and dimensions of the substrate do not get significantly altered. Thin film coating deposition processes are fundamental to many industries like cutting and forming tools, micro-tool fabrication, oil and gas, modern semiconductor manufacturing, and the fabrication of optical devices, such as compact disks, integrated circuits, solar photovoltaic cells, biocompatible coatings on implants, and so on. Some prominent applications include (i) protecting the mechanically functional substrates from wear and tear (ii) protecting the substrates from degradation through corrosion (iii) shielding the displays from environmental damage or avoiding scratches on them (iv) producing a precise level of reflectivity on semiconductor wafers, optical lenses, and create layers of metallization (iv) biocompatible coatings on implants. The development of thin film deposition methods has been significantly fuelled by the need for coatings with customized mechanical, physical, and chemical properties. The materials utilized, the kind of substrate, and the coating layer's thickness have significant impact on the effectiveness of such coatings. As technology advances, further research in thin film deposition is highly encouraged. This area of materials engineering is rapidly evolving, with ongoing discoveries and innovations driving the need for materials to support the next generation of technology. Thin film deposition remains a promising method to facilitate these advancements.

REFERENCES

1. Weinstein, Steven J., and Kenneth J. Ruschak. "Coating Flows." *Annual Review of Fluid Mechanics* 36 (2004): 29–53.
2. Krebs, Frederik C. "Fabrication and Processing of Polymer Solar Cells: A Review of Printing and Coating Techniques." *Solar Energy Materials and Solar Cells* 93, no. 4 (April 2009): 394–412.
3. Orudzhev, Farid, Shikhgasan Ramazanov, Dinara Sobola, Abdulgalim Isaev, Chuanyi Wang, Asiyat Magomedova, Makhmud Kadiev, and Kasinathan Kaviyarasu. "Atomic Layer Deposition of Mixed-Layered Aurivillius Phase on TiO_2 Nanotubes: Synthesis, Characterization and Photoelectrocatalytic Properties." *Nanomaterials* 10, no. 11 (2020): 2183.
4. Dallaev, R., D. Sobola, P. Tofel, L. Skvarenina, and P. Sedlak. "Aluminum Nitride Nanofilms by Atomic Layer Deposition Using Alternative Precursors Hydrazinium Chloride and Triisobutylaluminum." *Coatings* 10, no. 10 (2020): 954.
5. Konar, R., and G. A. Nessim. "A Mini-Review Focusing on Ambient-Pressure Chemical Vapor Deposition (AP-CVD) Based Synthesis of Layered Transition Metal Selenides for Energy Storage Applications." *Materials Advances* 3, no. 21 (2022): 4471–88.
6. Rossnagel, S. "Thin Film Deposition with Physical Vapor Deposition and Related Technologies." *Journal of Vacuum Science & Technology A* 21, no. S1 (2003): S74–S87.
7. Johnson, R., A. Hultqvist, and S. Bent. "A Brief Review of Atomic Layer Deposition: From Fundamentals to Applications." *Materials Today* 17, no. 5 (2014): 236–46.
8. Hong, Y.-S., S.-R. Lee, J.-H. Kim, and S.-Y. Lee. "Application of a DLC-Coating for Improving Hydrostatic Piston Shoe Bearing Performance under Mixed Friction Conditions." *International Journal of Precision Engineering and Manufacturing* 16, no. 2 (2015): 335–41.
9. Tandon, P., and H. Boek. "Experimental and Theoretical Studies of Flame Hydrolysis Deposition Process for Making Glasses for Optical Planar Devices." *Journal of Non-Crystalline Solids* 317 (2003): 275–89.
10. Ruano, J. M., V. Benoit, J. S. Aitchison, and J. M. Cooper. "Flame Hydrolysis Deposition of Glass on Silicon for the Integration of Optical and Microfluidic Devices." *Analytical Chemistry* 72, no. 6 (2000): 1093–97.
11. Yang, Y., K.-H. Kim, and J. Ong. "A Review on Calcium Phosphate Coatings Produced Using a Sputtering Process: An Alternative to Plasma Spraying." *Biomaterials* 26, no. 3 (2005): 327–41.

12. Mattox, D. M., ed. *Handbook of Physical Vapor Deposition (PVD) Processing, Film Formation, Adhesion, Surface Preparation and Contamination Control*. Westwood, NJ: Noyes Publications, 1998.
13. Glocker, D. B., and S. I. Shah, eds. *Handbook of Thin Film Process Technology*. Bristol: Institute of Physics Publishing, 1995.
14. Carlsson, J. O., and P. M. Martin. "Chemical Vapor Deposition." In *Handbook of Deposition Technologies for Films and Coatings*, edited by Peter M. Martin, 314–63. New York: William Andrew/Noyes, 2010.
15. Faraday, M. "Experimental Relations of Gold (and Other Metals) to Light." *Philosophical Transactions of the Royal Society of London* 147 (1857): 145–81.
16. Chander, S., and M. S. Dhaka. "Influence of Thickness on Physical Properties of Vacuum Evaporated Polycrystalline CdTe Thin Films for Solar Cell Applications." *Physica E: Low-Dimensional Systems and Nanostructures* 76 (2016): 52–59.
17. Von Pirani, M. "Production of Homogeneous Bodies from Tantalum or Other Metals." US Patent 848,600, filed 1907.
18. Rühle, R. "Verfahren zur Bedampfung im Vakuum." Deutsche Reich Patent 764,927, filed 1939.
19. Grove, W. R. "On the Electro-Chemical Polarity of Gases." *Philosophical Transactions of the Royal Society of London* 142 (1852): 87–101.
20. Kelly, P. J., and R. D. Arnell. "Magnetron Sputtering: A Review of Recent Developments and Applications." *Vacuum* 56, no. 3 (2000): 159–72.
21. Sproul, W. D., P. J. Rudnik, M. E. Graham, and S. L. Rohde. "High-Rate Reactive Sputtering in an Opposed Cathode Closed-Field Unbalanced Magnetron Sputtering System." *Surface and Coatings Technology* 43–44 (1990): 270–78.
22. Alami, J., S. Bolz, and K. Sarakinos. "High Power Pulsed Magnetron Sputtering: Fundamentals and Applications." *Journal of Alloys and Compounds* 483–84 (2009): 1–2.
23. Banerjee, T., and A. K. Chattopadhyay. "Structural, Mechanical and Tribological Properties of WS_2–Ti Composite Coating with and without Hard under Layer of TiN." *Surface and Coatings Technology* 258 (2014): 859–60.
24. Banerjee, T., and A. K. Chattopadhyay. "Structure, Mechanical and Tribological Characterizations of Pulsed DC Magnetron Sputtered TiN-WS_x Composite Coating." *Vacuum* 130 (2016): 93–104.
25. Belkind, A. "Characterization of Pulsed DC Magnetron Sputtering Plasmas." *New Journal of Physics* 7 (2005): 90.
26. Arnell, R. D., P. J. Kelly, and J. W. Bradley. "Recent Developments in Pulsed Magnetron Sputtering." *Surface and Coatings Technology* 188–89 (2004): 158–63.
27. Oliveira, J. C., F. Fernandes, F. Serra, and A. Cavaleiro. "On the Role of the Energetic Species in TiN Thin Film Growth by Reactive Deep Oscillation Magnetron Sputtering in Ar/N_2." *Thin Solid Films* 645 (2018): 253–64.
28. Oliveira, J. C., F. Fernandes, and A. Cavaleiro. "Tailoring the Nanostructure of Ti-Si-N Thin Films by HiPIMS in Deep Oscillation Magnetron Sputtering (DOMS) Mode." *Surface and Coatings Technology* 264 (2015): 140–49.
29. Kouznetsov, V., K. Macák, J. M. Schneider, U. Helmersson, and I. Petrov. "A Novel Pulsed Magnetron Sputter Technique Utilizing Very High Target Power Densities." *Surface and Coatings Technology* 122 (1999): 290–93.
30. Helmersson, U., M. Lattemann, J. Bohlmark, A. P. Ehiasarian, and J. T. Gudmundsson. "Ionized Physical Vapor Deposition (IPVD): A Review of Technology and Applications." *Thin Solid Films* 513, no. 1 (2006): 1–24.
31. Anders, A. "Discharge Physics of High Power Impulse Magnetron Sputtering." *Surface and Coatings Technology* 205 (2011): S1–S9.
32. Bohlmark, J., M. Lattemann, J. T. Gudmundsson, A. P. Ehiasarian, Y. A. Gonzalvo, N. Brenning, and U. Helmersson. "The Ion Energy Distributions and Plasma Composition

of a High Power Impulse Magnetron Sputtering Discharge." *Thin Solid Films* 515 (2006): 1522–26.

33. Samuelsson, M., D. Lundin, J. Jensen, M. A. Raadu, J. T. Gudmundsson, and U. Helmersson. "On the Film Density Using High Power Impulse Magnetron Sputtering." *Surface and Coatings Technology* 205 (2010): 591–96.
34. Sittinger, V., G. Ruske, W. Werner, C. Jacobs, B. Szyszka, and D. Christie. "High Power Pulsed Magnetron Sputtering of Transparent Conducting Oxides." *Thin Solid Films* 516 (2008): 5847–59.
35. Lattemann, M., U. Helmersson, and J. E. Greene. "Fully Dense, Nonfaceted 111-Textured High Power Impulse Magnetron Sputtering TiN Films Grown in the Absence of Substrate Heating and Bias." *Thin Solid Films* 518 (2010): 5978–80.
36. Konstantinidis, S., J.-P. Dauchot, and M. Hecq. "Titanium Oxide Thin Films Deposited by High-Power Impulse Magnetron Sputtering." *Thin Solid Films* 515 (2006): 1182–86.
37. Alami, J., P. Eklund, J. M. Andersson, M. Lattemann, E. Wallin, J. Bohlmark, P. Persson, and U. Helmersson. "Phase Tailoring of Ta Thin Films by Highly Ionized Pulsed Magnetron Sputtering." *Thin Solid Films* 515 (2007): 3434–38.
38. Paulitsch, J., P. H. Mayrhofer, C. Mitterer, W.-D. Münz, and M. Schenkel. "Mechanical and Tribological Properties of CrN Coatings Deposited by a Simultaneous HIPIMS/UBM Sputtering Process." In *Society of Vacuum Coaters 50th Annual Technical Conference Proceedings*, Louisville, KY, April 28–May 3, 2007, 150.
39. Ehiasarian, A. P., J. G. Wen, and I. Petrov. "Interface Microstructure Engineering by High Power Impulse Magnetron Sputtering for the Enhancement of Adhesion." *Journal of Applied Physics* 101 (2007): 054301.
40. Alami, J., and P. O. Å. Persson. "Ion-Assisted Physical Vapor Deposition for Enhanced Film Properties on Nonflat Surfaces." *Journal of Vacuum Science & Technology A* 23, no. 2 (2005): 278–80.
41. Aijaz, A., D. Lundin, P. Larsson, and U. Helmersson. "Dual-Magnetron Open Field Sputtering System for Sideways Deposition of Thin Films." *Surface and Coatings Technology* 204 (2010): 2165–69.
42. Wallin, E., T. I. Selinder, M. Elfwing, and U. Helmersson. "Synthesis of α-Al_2O_3 Thin Films Using Reactive High-Power Impulse Magnetron Sputtering." *Europhysics Letters* 82 (2008): 36002.
43. Ritala, M., J. Niinistö, S. Krumdieck, P. Chalker, H. Pemble, W. L. Gladfelter, B. Leese, R. A. Fischer, H. Parala, R. D. Dupuis, S. E. Alexandrov, S. J. C. Irvine, R. Palgrave, and I. P. Parkin. "Chemical Vapour Deposition." In *Chemical Vapour Deposition*, edited by A. C. Jones and M. L. Hitchman. Cambridge: The Royal Society of Chemistry, 2009.
44. Manasevit, H. M. "Single-Crystal Gallium Arsenide on Insulating Substrates." *Applied Physics Letters* 12 (1968): 156–58.
45. Manasevit, H. M. "The Use of Metalorganics in the Preparation of Semiconductor Materials: Growth on Insulating Substrates." *Journal of Crystal Growth* 13–14 (1972): 306–14.
46. Manasevit, H. M., and W. I. Simpson. "The Use of Metal-Organics in the Preparation of Semiconductor Materials: I. Epitaxial Gallium-V Compounds." *Journal of the Electrochemical Society* 116 (1969): 1725.
47. Suntola, T. "Atomic Layer Epitaxy." *Materials Science Reports* 4 (1989): 261–312.
48. Kurek, A., P. G. Gordon, S. Karle, A. Devi, and S. T. Barry. "Recent Advances Using Guanidinate Ligands for Chemical Vapour Deposition (CVD) and Atomic Layer Deposition (ALD) Applications." *Australian Journal of Chemistry* 67, no. 8 (2014): 989–96.
49. Gugua, B. C., C. O. Ujah, C. O. Asadu, D. V. V. Kallon, and B. N. Ekwueme. "Electroplating in the Modern Era, Improvements and Challenges: A Review." *Hybrid Advances* 7 (2024): 100286.
50. Mallory, O., and J. B. Hajdu. *Electroless Plating: Fundamentals and Applications*. Orlando, FL: American Electroplaters and Surface Finishers Society, 1990.

51. Okinaka, Y., and T. Osaka. "Electroless Deposition Processes: Fundamentals and Applications." In *Advances in Electrochemical Science and Engineering*, edited by H. Gerischer and C. V. Tobias, 55. New York: VCH Publishers, 1994.
52. Djokic, S. S. "Electroless Deposition of Metals and Alloys." In *Modern Aspects of Electrochemistry*, edited by B. E. Conway and R. E. White, 51. New York: Kluwer Academic/Plenum Publishers, 2002.
53. Banerjee, T., S. S. Sen, B. Oraon, and G. Majumdar. "Predicting Electroless Ni-Co-P Coating Using Response Surface Method." *International Journal of Advanced Manufacturing Technology* 64 (2013): 1729–36.
54. Kumar, S., T. Banerjee, and D. Patel. "Structural, Mechanical, Tribological and Corrosion Characteristics of Dual Layer Ni-P/Sr-P Coating on AZ91D Mg Alloy through Environmentally-Friendly Chemical Conversion and Electroless Routes." *Surface and Coatings Technology* 485 (2024): 130922.
55. Kumar, S., T. Banerjee, and D. Patel. "Synthesis and Evaluation of Tribo-Mechanical and Corrosion Properties of Electroless Ni-P Coating on AZ91D Mg Alloy with Stannate Conversion Underlayer." *Proceedings of the Institution of Mechanical Engineers, Part L: Journal of Materials: Design and Applications* (2024): 1–17. https://doi.org/10.1177/14644207241300643.
56. Boipai, B., and T. Banerjee. "Study on Adhesion Failure of Electrically Conducting Electroless Ni-P Coating on Polyethylene Terephthalate Substrate." *Materials Today: Proceedings* 66 (2022): 3930–34.
57. Emslie, A., F. Bonner, and L. Peck. "Flow of a Viscous Liquid on a Rotating Disk." *Journal of Applied Physics* 29, no. 6 (1958): 858–64.
58. Scriven, L. "Physics and Applications of DIP Coating and Spin Coating." *MRS Proceedings* 121 (1988): 717–29.
59. Puetz, J., and M. A. Aegerter. "Dip Coating Technique." In *Sol-Gel Technologies for Glass Producers and Users*, edited by M. A. Aegerter and M. Mennig, 37–48. Boston: Springer, 2004.

2 Advanced Deposition Techniques and Industrial Applications of Thin Films

Venkata Sushma Chinta, Ashutosh Sahu, Amireddy Kiran Kumar, and Lavish Kumar Singh

2.1 INTRODUCTION

In contemporary science and technology, there is a significant focus on reducing materials to nano-scale dimensions, including both nano-size and nano-thickness. This miniaturization results in unique properties in areas such as electrical, optoelectronic, optical, and dielectric applications. Consequently, a subfield of materials science called coatings or thin films has emerged. A thin film is characterized as a material layer with a thickness spanning from several nanometres to several micrometres. Like bulk materials, thin films may display either amorphous or polycrystalline structures, contingent upon the material's characteristics and the preparation circumstances. Thin films are composed of two primary elements: the film layer and the substrate that supports the deposition of the layer. Electrochromic cells and thin-film solar cells are two examples of applications that make use of these films' multi-layer construction. Physical and chemical deposition are the two most frequent methods used to create high-quality thin films. The categorization of deposition methods is illustrated in Figure 2.1.

2.2 PHYSICAL DEPOSITION TECHNIQUES (PDTs)

Evaporation techniques are widely used for depositing materials as thin films. These techniques have as their fundamental idea transforming the material from its solid state to a vapour state and then back to a solid state on a designated substrate. This entire process occurs under controlled or vacuum atmospheric conditions.

2.2.1 Vacuum Thermal Evaporation Technique (VTET)

The VTET stands out as one of the simplest techniques for the preparation of amorphous thin films, especially chalcogenide films like CdSSe [1, 2], Ge-Te-Ga [3, 4], and MnS [5, 6]. Chalcogenide materials exhibit remarkable versatility and are utilized

DOI: 10.1201/9781003750659-2

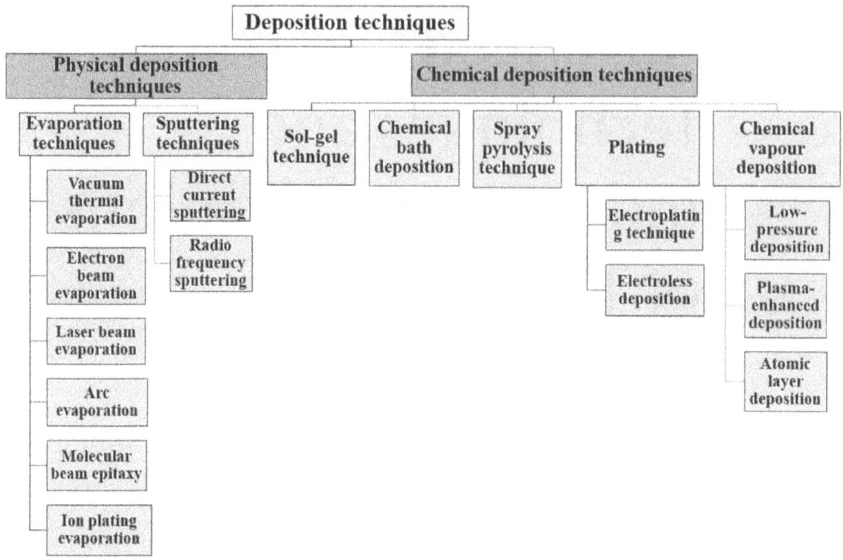

FIGURE 2.1 Thin film deposition techniques.

across diverse domains owing to their adjustable optical and electrical characteristics [7]. They are integral to memory-switching devices [8, 9], phase-change materials [10], and solar energy technologies [11]. Their adjustable bandgap and excellent optical characteristics make them ideal for photonics and optoelectronic devices, such as infrared detectors and modulators. Additionally, chalcogenides like CdSxSe1-x are essential in photovoltaic cells, LEDs, infrared optics, memory devices, and sensors for environmental monitoring and biological detection. Moreover, ongoing developments in deposition methods are progressively refining the structural and functional characteristics of these films, thereby boosting their efficacy in innovative technologies.

The VTET is a distinct form of vacuum evaporation and is fundamentally dependent on two critical factors: the material undergoing thermal vaporization and the application of a potential difference to the substrate in a vacuum setting, usually between 10^{-5} mbar and 10^{-9} mbar. This method is necessary to make sure that the thin films are deposited correctly and are of good quality. Figure 2.2 presents the schematic configuration for thermal evaporation. The assembly features a substrate holder that is affixed to a planetary rotation mechanism, which is strategically located directly above the evaporation source.

2.2.2 Electron Beam Evaporation (EBE)

EBE is a physical deposition technique that involves generating a concentrated electron beam from a filament. This beam is directed through electric and magnetic fields to impact the target material, resulting in its vaporization within a vacuum environment, as illustrated in Figure 2.3. Thin films produced by this technique are

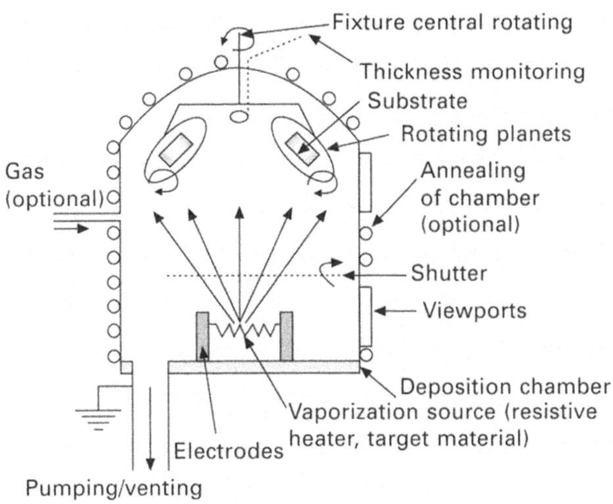

FIGURE 2.2 Thermal deposition techniques.

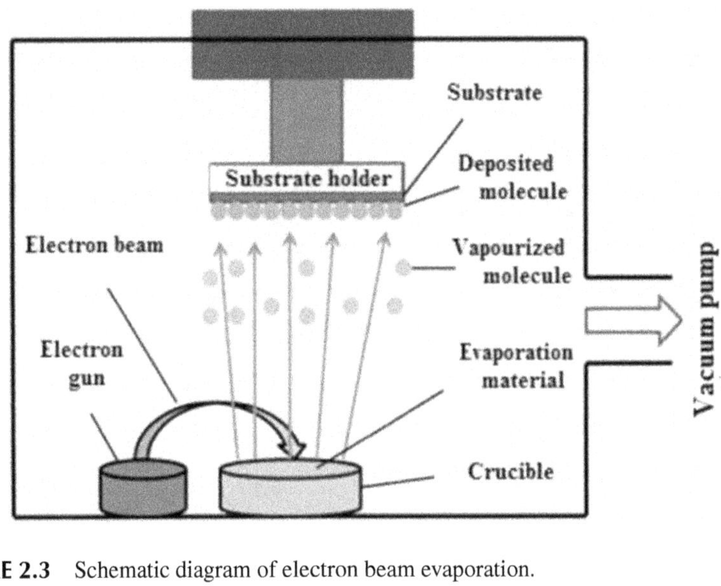

FIGURE 2.3 Schematic diagram of electron beam evaporation.

known for their high purity and quality [12–15]. A wide range of materials can be prepared using EBE, including metals [16], oxides [17], crystalline and amorphous semiconductors [18], and molecular materials [19]. EBE is a highly adaptable method extensively used for creating high-precision optical coatings, advanced microelectronics, and photovoltaic cells.

In photovoltaics, EBE is advantageous for manufacturing efficient thin-film solar cells, like Cu_2SnS_3 (CTS) cells, due to its high-purity and precise deposition capabilities [20]. It also excels in producing specialized coatings for protective and functional uses across various industries. EBE's ability to achieve high growth rates and provide conformal coatings on structured surfaces enhances the performance and cost-effectiveness of silicon-based devices [21].

2.2.3 Pulsed Laser Deposition (PLD)

PLD is a commonly employed physical deposition method for the fabrication of thin-film coatings. [22]. As seen in Figure 2.4, in this technique, a laser beam ablates a target material inside a vacuum chamber. Various types of lasers can be employed for this purpose, with the XeCl (308 nm), KrF (248 nm), and Nd-YAG laser being the most common choices. The neutral atoms, ionized species, and ground-state atoms produced by the laser beam striking the target material deposit onto different substrates. In the process of depositing metal oxide thin films, the introduction of oxygen facilitates the formation of metal oxides [23, 24]. The characteristics of thin films generated through pulsed laser deposition are affected by various parameters, such as the ambient gas pressure, wavelength, energy of the laser, pulse duration, and the separation between the target and the substrate [25].

Techniques such as molecular isotopic spectroscopy with laser ablation [26], optical emission spectrum analysis [27], and laser-induced fluorescence [28] can be used to control and monitor the ablation process. The temperature of the substrate plays a vital role in influencing the morphology of the thin films that are deposited. PLD facilitates thin film deposition through three primary growth modes: Stranski–Krastanov, Frank–van der Merwe, and Volmer–Weber [29]. In comparison to alternative physical deposition techniques, PLD presents benefits including swift deposition rates and the ability to work with both oxygen and various inert gases. PLD has

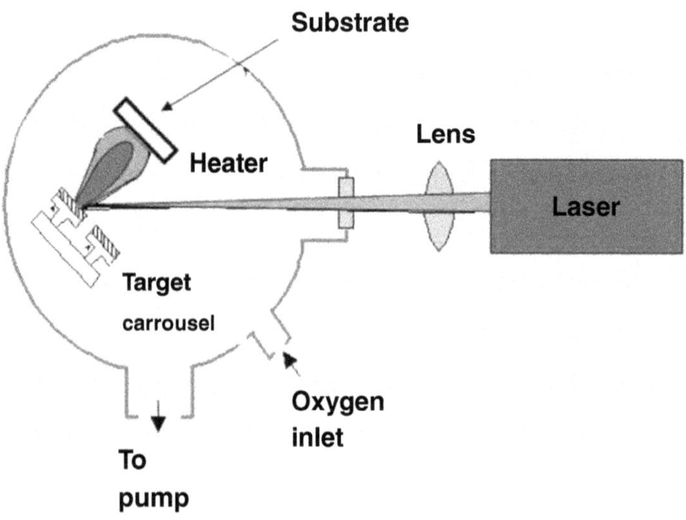

FIGURE 2.4 Schematic diagram of pulsed laser deposition.

been adeptly employed in a range of applications, encompassing the fabrication of high-temperature superconductors, ferroelectric thin films for electronic devices, and protective coatings for materials subjected to extreme conditions. Its versatility and precision make it a valuable tool in the development of advanced materials and technologies [30].

2.3 SPUTTERING TECHNIQUE (ST)

The ST is primarily utilized for the deposition of metal and oxide films, enabling meticulous control over the surface roughness and crystalline structure. A sputtering system fundamentally consists of an evacuated chamber featuring a metallic anode and cathode, which enables a glow discharge within the residual gas of the chamber. Utilizing a voltage within several keV and sustaining a pressure exceeding 0.01 mbar facilitate the process of film deposition [31]. This process depends on ions from the discharge impacting the cathode, leading to the ejection of molecules from the cathode with significant kinetic energy. To achieve optimal momentum transfer, it is essential that the atomic weight of the bombarding ions closely matches that of the target material. The ejected molecules move in linear trajectories, impacting the anode or substrate to create a compact thin film. Figure 2.5 presents a schematic representation of the sputtering system.

Sputtering presents a range of benefits. This process facilitates the straightforward synthesis of high-melting-point materials and yields films with compositions that closely resemble those of the initial materials. This method is compatible with reactive gases, such as oxygen, and is appropriate for ultrahigh vacuum applications. However, it has limitations, such as the inability to produce thick coatings and the challenge of achieving uniform deposition on complex shapes. There are two primary types of ST namely radio frequency (RF) sputtering and direct current (DC)

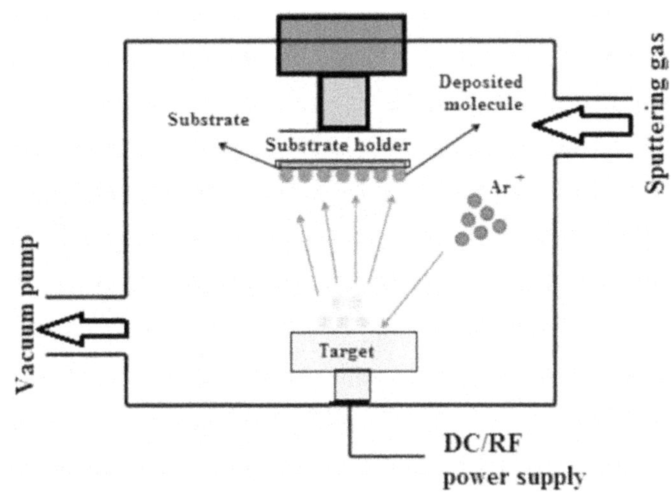

FIGURE 2.5 Sputtering system.

sputtering [32]. RF sputtering utilizes RF power and is commonly used for most dielectric materials. DC sputtering on the other hand, which uses DC power, is typically employed with electrically conductive target materials due to its ease of control and cost-effectiveness. Aluminium nitride films, for example, have been prepared using both DC and RF sputtering techniques, and their structural and optical properties have been compared [33].

Applications of sputtering are diverse and significant in various fields. In the semiconductor sector, it serves the purpose of fabricating thin films for integrated circuits and various electronic devices. Sputtered films are also crucial in the production of solar cells, where they help in forming the photovoltaic layers that convert sunlight into electricity. Additionally, sputtering is employed in the creation of optical coatings for lenses and mirrors, providing anti-reflective or reflective properties. In the automotive industry, sputtered coatings are used to enhance the durability and performance of various components. Furthermore, sputtering is applied in the data storage industry for creating magnetic films for hard drives and in the decorative coating industry for producing aesthetically pleasing finishes on consumer products. The versatility and precision of sputtering make it an invaluable technique in modern manufacturing and technology [31].

2.4 CHEMICAL DEPOSITION TECHNIQUES (CDTs)

While physical methods for producing thin films yield high-quality results and functional properties, they are often costly and require substantial amounts of target materials. In light of the need for cost-effective production of high-quality thin films, there has been a significant adoption of CDT. These methods are cost-effective and capable of producing good-quality films without the need for expensive equipment. CDTs are significantly influenced by the chemistry of solutions, including factors such as viscosity and pH value. Among these techniques, the spray pyrolysis, chemical bath deposition, chemical vapour deposition, electrodeposition, and sol-gel process are the most commonly used operations.

2.4.1 Sol-Gel Technique (SGT)

The SGT is a versatile and cost-effective method for synthesizing thin films and powders and is known for its low-temperature processing and precise control over film composition and microstructure. SGT involves the controlled polycondensation and hydrolysis of metal alkoxides (e.g., Si, Ti, Zr, V, Al, Mo, W, Sn, Ge, and Zn) or chlorides in a liquid medium to produce a colloidal solution, or sol, composed of suspended nanoparticles [34, 35]. The resulting sol is subsequently deposited onto substrates utilizing techniques like dip coating and spin coating, as illustrated in Figure 2.6. Dip coating is commonly utilized to produce transparent oxide films on substrates with high surface quality and planarity, allowing for accurate control of film thicknesses up to 1 μm. The procedure consists of five distinct stages: dipping, initialization, film formation, runoff, and evaporation. Spin coating, on the other hand, involves dripping the solution onto a spinning substrate to spread it evenly and form a uniform film, making it suitable for smaller substrates like disks or lenses [36].

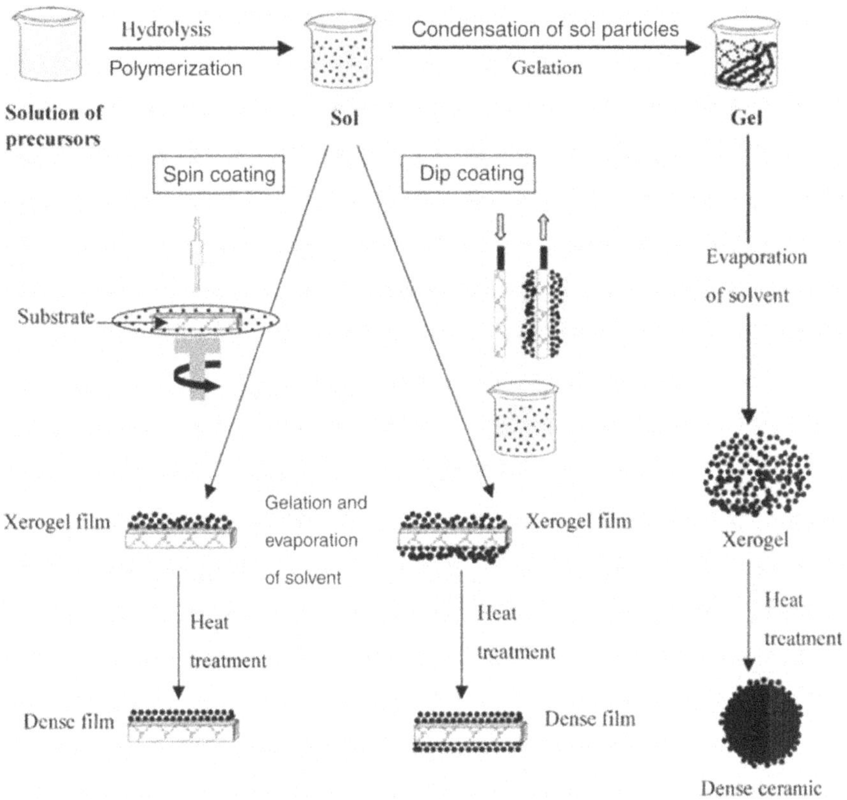

FIGURE 2.6 Schematic diagram of sol-gel technique [36].

After deposition, the sol undergoes a heat treatment process to form a xerogel, which is a dried gel at ambient pressure. The choice of drying method – such as supercritical drying or freeze-drying – depends on the desired final product. These methods can produce various materials, including xerogels, ambigels, cryogels, and aerogels, each with specific properties like high surface area and nanoscale dimensions.

SGT finds wide-ranging applications in optical coatings [37–39], where precise control over properties like reflectance, transmittance, and refractive index is crucial. Additionally, sol-gel films play a vital role in electronics, enabling advancements in blue laser diodes, LEDs, and optically controlled switches. Moreover, sol-gel films are instrumental in protective coatings, safeguarding metals from corrosion and oxidation, especially in high-temperature environments. Titanium dioxide (TiO_2) coatings, for instance, exhibit photo-induced charge transfer properties useful in pollutant degradation and solar cells. Porous films synthesized through the SGT process offer high surface areas, making them ideal for applications such as gas and chemical sensing, solar cells, and surface reactions. The versatility and adaptability of sol-gel coatings make them indispensable in modern technology and manufacturing processes.

2.4.2 Chemical Bath Deposition Technique (CBDT)

CBDT, recognized as controlled precipitation, stands as one of the most established techniques for film deposition on substrates. This method is mainly employed for the preparation of chalcogenide and metal oxide films, typically at reduced temperatures. The CBDT method involves a precursor solution where metal ions are complexed with various ligands, including ammonia, triethanolamine, ethylenediaminetetraacetic acid (EDTA), or citric acid. Once complexation is complete, anions are introduced from sources like thiourea, thioacetamide, thiosulphate, or sodium sulphide for sulphur, and selenourea or sodium selenosulphate for selenium to form chalcogenides [40]. The substrates are positioned in a specified orientation horizontally or vertically within the solution until the target film thickness is attained. The procedure for oxide film formation exhibits minor variations; subsequent to complexation and pH adjustment, the substrate is submerged in the solution and subjected to heating within a range of 60 °C to 100 °C to facilitate the deposition of metal hydroxide films, which may subsequently be transformed into oxides via annealing. The experimental set-up for CBDT typically involves stationary substrates and solutions (see Figure 2.7). A magnetic stirrer is usually used to stir the solution. The chemical bath is heated to the required temperature using water or paraffin baths, with stirring

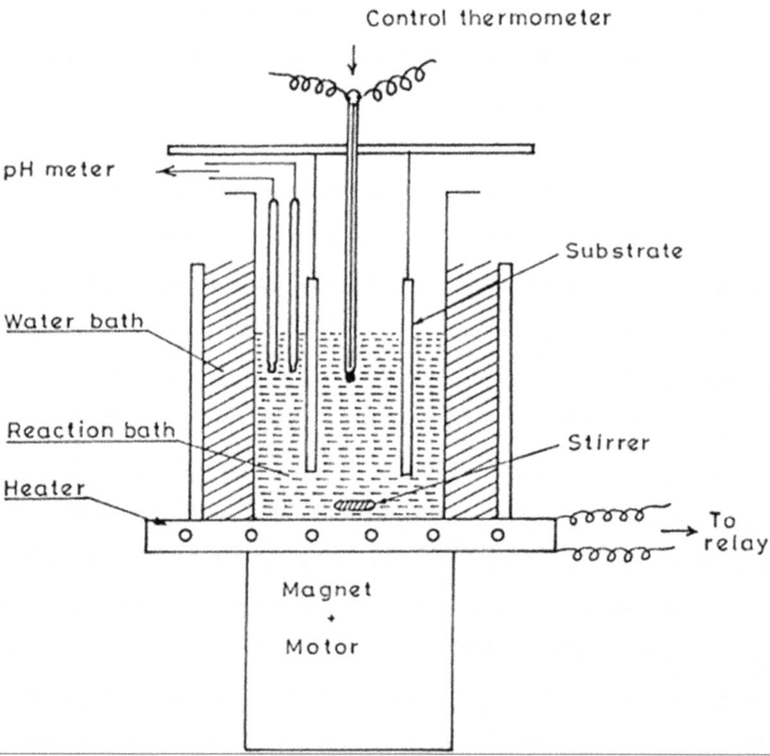

FIGURE 2.7 Chemical bath deposition technique [40].

either starting from room temperature or beginning after the desired temperature is reached. This approach facilitates accurate regulation of film thickness and composition, rendering it a flexible technique suitable for a range of applications.

CBDT offers a versatile approach for depositing thin layers with varied applications. Using chemical deposition methods, a large number of films such as CdS, ZnS, Ag_2S, As_2S_3, SnS, In_2S_3, CuS, NiS, NiSe, and so on have been deposited. Polycrystalline CdS films [41], crucial for photoconductors and photovoltaic devices, are synthesized alongside ZnS [42] films, widely used in photoluminescent devices and diodes. Ag_2S [43] thin films find utility in various technologies, including solar cells and IR detectors. Antimony trisulphide films cater to microwave devices and optoelectronic applications. As_2S_3 [44] exhibits photoinduced effects essential for hologram recording and circuit fabrication. SnS [45] holds promise for solar energy conversion, while In_2S_3 [46] serves as a buffer layer in solar cells. PbS [47] is essential for infrared detectors and solar control coatings. Cu_xS [48] films, notable for their solar control characteristics, are applied in diverse devices. Lastly, NiS and [49] NiSe films are utilized in IR detectors and photovoltaic devices, respectively, showcasing the broad applicability of CBDT in thin film technology.

2.4.3 Spray Pyrolysis Technique (SPT)

The process of SPT entails the deposition of thin films via the atomization of a precursor solution onto a heated substrate, as illustrated in Figure 2.8. The process begins with the preparation of a precursor solution through the dissolution of the desired materials in an appropriate solvent, ensuring thorough mixing. This solution is then loaded into an atomizer, which generates fine droplets; the atomizer's settings are adjusted based on the required droplet size and distribution. The substrate, placed on a heater, is maintained at a specific temperature controlled by a temperature controller [50]. Upon contact with the heated substrate, the atomized droplets undergo solvent evaporation, resulting in the formation of a thin film. The deposition rate and

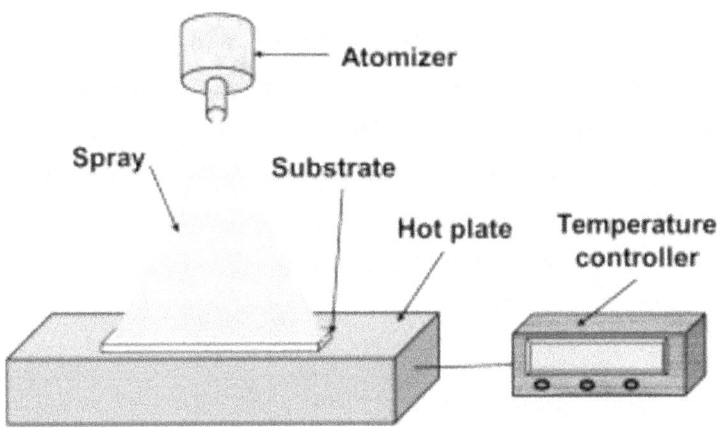

FIGURE 2.8 Spray pyrolysis technique [51].

duration are managed to achieve the desired film thickness. Post-deposition, the film is allowed to cool and may undergo annealing to enhance its properties. Finally, the thin film is analysed using methods like scanning electron microscopy, X-ray diffraction, and various spectroscopic methods to evaluate its thickness, morphology, crystal structure, and other relevant properties. This method, despite challenges like thin film quality and vapour convection, is cost-effective and scalable, making it suitable for applications in electronics, optoelectronics, and photovoltaics [51].

2.4.4 Plating Technique (PT)

Electroplating, also known as electrodeposition, is a vital technique for creating thin films, which play a significant role in various technological areas. This process involves passing an electrical current through a solution containing dissolved metal ions and the object to be plated. The typical set-up includes an anode (positively charged electrode), which serves as the source of the material to be deposited, and a cathode (negatively charged electrode), which is the substrate to be coated with a thin film (see Figure 2.9) [52, 53]. The electroplating process occurs in a plating bath, usually a non-metallic tank filled with an electrolyte containing the metal ions to be deposited. The anode, often made of the metal intended for deposition, is connected to the positive terminal of a power supply, while the cathode, the substrate for the thin film, is connected to the negative terminal. A stable power supply with minimal

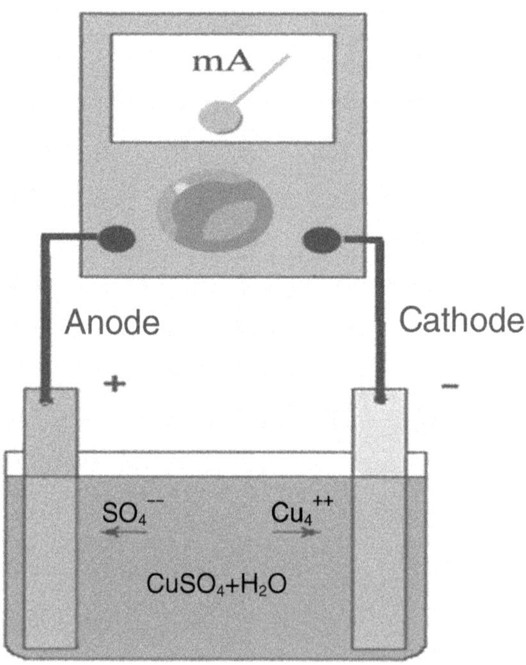

FIGURE 2.9 Electroplating [54].

ripple noise is essential to provide a steady bias current that remains constant despite changing loads during deposition.

Upon the application of direct current between the anode and cathode, positive metal ions present in the solution move towards the negatively charged cathode, resulting in their precipitation and the formation of a thin film on its surface. To maintain ionic balance, the metal from the anode dissolves into the solution, replenishing the deposited ions. In specific instances, like gold plating, the anode is constructed from a material that remains inert in the electrolyte, such as titanium. The deposition of gold from the solution occurs via an oxidation–reduction reaction, in which one material undergoes oxidation by losing electrons, while another material is reduced by gaining electrons. Different metals and alloys can be electroplated as thin films depending on the solution and conditions. Metals like copper, nickel, and zinc can be deposited from aqueous solutions. Elements like aluminium and titanium require organic electrolytes, while metals such as niobium, magnesium, tungsten, and tantalum are plated from molten salt electrolytes at high temperatures (700 °C and above).

Electroplated thin films are essential in various fields. In electronics, copper and gold thin films enhance conductivity in circuits and devices. Optoelectronic applications use silver and aluminium for reflective coatings and indium titanium oxide (ITO) coating for transparent conductive films in displays. Surface modifications benefit from nickel and chromium for corrosion resistance and titanium nitride for wear resistance. In medical devices, gold, platinum, and titanium improve biocompatibility, while electroplated coatings ensure sterilization resistance. Energy storage applications use lithium thin films in batteries and platinum in fuel cells. Decorative applications include enhancing jewellery and consumer goods with precious metal coatings.

2.5 CHEMICAL VAPOUR DEPOSITION (CVD)

CVD is a process that involves the employment of chemical reactions on a heated substrate to deposit solid materials from a vapour phase [55, 56]. This method produces thin films, powders, or single crystals. By manipulating variables such as the substrate type, temperature, gas mixture, and pressure, CVD can create materials with various physical and chemical properties. Known for its ability to create uniform coatings with low porosity, CVD is utilized in numerous applications, including the fabrication of passivation layers, conductors, oxidation barriers, and dielectrics [57, 58]. It is also employed in producing carbon nanotubes solar cells and high-temperature materials without the need for post-deposition annealing.

Plasma-enhanced chemical vapour deposition (PECVD) represents a modification of the CVD method, employing plasma to provide a portion of the energy necessary for the chemical reactions that facilitate the deposition of material onto a substrate. The plasma produced through a radio frequency discharge generates reactive radicals and ions from the precursor gas, facilitating the deposition process at reduced substrate temperatures in comparison to traditional CVD techniques (see Figure 2.10). This characteristic allows for higher deposition rates and well-adherent films without damaging temperature-sensitive underlying layers.

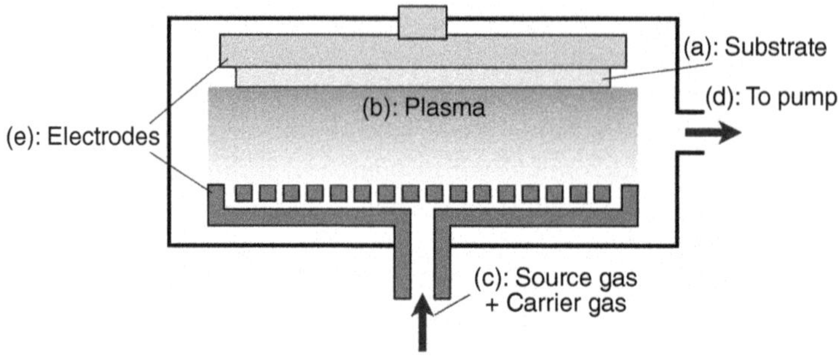

FIGURE 2.10 Plasma-enhanced chemical vapour deposition [59].

The energy from the plasma boosts the chemical activity of the gas mixture, aiding in the creation of films that exhibit favourable characteristics, including consistent thickness and robust adhesion to the substrate. By operating at reduced temperatures, PECVD is particularly advantageous for the final layers of multi-layered devices, where high temperatures could compromise the integrity of previously deposited layers. PECVD is extensively used in the semiconductor industry for depositing insulating layers, passivation films, and barrier layers. It is also employed in producing coatings for various applications, including optical films, wear-resistant surfaces, and protective coatings [59, 60]. The capability to manipulate film characteristics through the modification of plasma parameters, such as power and gas flow rates, makes PECVD a versatile and essential tool in advanced material fabrication and surface engineering.

2.6 SUMMARY

Thin film deposition is a pivotal process in modern technology, enabling the development of advanced materials with specific properties for various industrial applications such as microelectronics, healthcare, decorative arts, and optics. This chapter highlighted both physical and chemical deposition techniques. Physical methods, such as evaporation, sputtering, and pulsed laser deposition, deliver high-quality films with excellent functional properties, though they often incur high costs and require substantial target materials. On the other hand, chemical methods, including the chemical bath deposition, sol-gel process, spray pyrolysis, electroplating, and chemical vapour deposition, provide cost-effective alternatives that can produce good-quality films using simpler, less expensive equipment. Among these, the sol-gel technique stands out for its versatility and ability to create films with precise composition and microstructure control at low temperatures. CBDT is notable for its ability to deposit various chalcogenide and metal oxide films at relatively low temperatures. Spray pyrolysis and plating techniques offer scalable, cost-effective solutions for electronic and optoelectronic applications. CVD and its variant, PECVD, enable the formation of uniform, adherent films essential for semiconductor devices and protective coatings.

REFERENCES

1. Hassanien, A. S., and A. A. Akl. "Effect of Se Addition on Optical and Electrical Properties of Chalcogenide CdSSe Thin Films." *Superlattices and Microstructures* 89 (2016): 153–169.
2. Liu, Y., J. A. Zapien, Y. Y. Shan, C.-Y. Geng, C. S. Lee, and S.-T. Lee. "Wavelength-Controlled Lasing in ZnxCd1-xS Single-Crystal Nanoribbons." *Advanced Materials* 17, no. 11 (2005): 1372–1377.
3. Wang, G., Q. Nie, X. Shen, F. Chen, J. Li, W. Zhang, and S. Dai. "Phase Change and Optical Band Gap Behavior of Ge–Te–Ga Thin Films Prepared by Thermal Evaporation." *Vacuum* 86, no. 10 (2012): 1572–1575.
4. Lencer, D., M. Salinga, B. Grabowski, T. Hickel, J. Neugebauer, and M. Wuttig. "A Map for Phase-Change Materials." *Nature Materials* 7, no. 12 (2008): 972–977.
5. Hannachi, A., A. Segura, and H. Maghraoui-Meherzi. "Growth of Manganese Sulfide (α-MnS) Thin Films by Thermal Vacuum Evaporation: Structural, Morphological and Optical Properties." *Materials Chemistry and Physics* 181 (2016): 326–332.
6. Gümüş, C., C. Ulutaş, R. Esen, O. M. Özkendir, and Y. Ufuktepe. "Preparation and Characterization of Crystalline MnS Thin Films by Chemical Bath Deposition." *Thin Solid Films* 492, no. 1–2 (2005): 1–5.
7. Frumar, M., B. Frumarova, and T. Wagner. "Amorphous and Glassy Semiconducting Chalcogenides." In *Comprehensive Semiconductor Science and Technology*, 206–261. 2011. Elsevier Science.
8. Malligavathy, M., R. T. Ananth Kumar, C. Das, S. Asokan, and D. Pathinettam Padiyan. "Growth and Characteristics of Amorphous Sb_2Se_3 Thin Films of Various Thicknesses for Memory-Switching Applications." *Journal of Non-Crystalline Solids* 429 (2015): 93–97.
9. Ananth Kumar, R. T., C. Das, P. Chithra Lekha, S. Asokan, C. Sanjeeviraja, and D. Pathinettam Padiyan. "Enhancement in Threshold Voltage with Thickness in Memory Switch Fabricated Using $GeSe_{1.5}S_{0.5}$ Thin Films." *Journal of Alloys and Compounds* 615 (2014): 629–635.
10. Sangeetha, B. G., C. M. Joseph, and K. Suresh. "Preparation and Characterization of $Ge_1Sb_2Te_4$ Thin Films for Phase-Change Memory Applications." *Microelectronic Engineering* 127 (2014): 77–80.
11. Salomé, P. M. P., H. Rodriguez-Alvarez, and S. Sadewasser. "Incorporation of Alkali Metals in Chalcogenide Solar Cells." *Solar Energy Materials and Solar Cells* 143 (2015): 9–20.
12. Orava, J., T. Kohoutek, and T. Wagner. "Deposition Techniques for Chalcogenide Thin Films." In *Chalcogenide Glasses*, 265–309. Elsevier Science & Technology, Elsevier Ltd., 2014.
13. Lokhande, A. C., R. B. V. Chalapathy, M. He, E. Jo, M. Gang, S. A. Pawar, and J. H. Kim. "Development of Cu_2SnS_3 (CTS) Thin Film Solar Cells by Physical Techniques: A Status Review." *Solar Energy Materials and Solar Cells* 153 (2016): 84–107.
14. Tang, Z., K. Kosaka, H. Uegaki, J. Chantana, Y. Nukui, D. Hironiwa, and T. Minemoto. "Investigation on Evaporation and Suppression of SnS during Fabrication of Cu_2SnS_3 Thin Films." *Physica Status Solidi (a)* 212, no. 10 (2015): 2289–2296.
15. Barranco, A., A. Borras, A. R. Gonzalez-Elipe, and A. Palmero. "Perspectives on Oblique Angle Deposition of Thin Films: From Fundamentals to Devices." *Progress in Materials Science* 76 (2016): 59–153.
16. Mukherjee, S., and D. Gall. "Structure Zone Model for Extreme Shadowing Conditions." *Thin Solid Films* 527 (2013): 158–163.
17. Schulz, U., S. G. Terry, and C. G. Levi. "Microstructure and Texture of EB-PVD TBCs Grown under Different Rotation Modes." *Materials Science and Engineering: A* 360, no. 1–2 (2003): 319–329.

18. Merkel, J. J., T. Sontheimer, B. Rech, and C. Becker. "Directional Growth and Crystallization of Silicon Thin Films Prepared by Electron-Beam Evaporation on Oblique and Textured Surfaces." *Journal of Crystal Growth* 367 (2013): 126–130.
19. Yang, B., H. Duan, C. Zhou, Y. Gao, and J. Yang. "Ordered Nanocolumn-Array Organic Semiconductor Thin Films with Controllable Molecular Orientation." *Applied Surface Science* 286 (2013): 104–108.
20. Shah, A. V., H. Schade, M. Vanecek, J. Meier, E. Vallat-Sauvain, N. Wyrsch, and J. Bailat. "Thin-Film Silicon Solar Cell Technology." *Progress in Photovoltaics: Research and Applications* 12, no. 23 (2004): 113–142.
21. Matsuyama, T., N. Terada, T. Baba, T. Sawada, S. Tsuge, K. Wakisaka, and S. Tsuda. "High-Quality Polycrystalline Silicon Thin Film Prepared by a Solid Phase Crystallization Method." *Journal of Non-Crystalline Solids* 198–200 (1996): 940–944.
22. Lorenz, M., and M. S. Ramachandra Rao. "25 Years of Pulsed Laser Deposition." *Journal of Physics D: Applied Physics* 47, no. 3 (2013): 030301.
23. Shepelin, N. A., Tehrani, Z. P., Ohannessian, N., Schneider, C. W., Pergolesi, D., & Lippert, T. " A Practical Guide to Pulsed Laser Deposition." *Chemical Society Reviews* 52, no. 7 (2023): 2294–2321.
24. Lowndes, D. H., D. B. Geohegan, A. A. Puretzky, D. P. Norton, and C. M. Rouleau. "Synthesis of Novel Thin-Film Materials by Pulsed Laser Deposition." *Science* 273, no. 5277 (1996): 898–903.
25. Ashfold, M. N. R., F. Claeyssens, G. M. Fuge, and S. J. Henley. "Pulsed Laser Ablation and Deposition of Thin Films." *Chemical Society Reviews* 33, no. 1 (2004): 23–38.
26. Russo, R. E., X. Mao, J. J. Gonzalez, V. Zorba, and J. Yoo. "Laser Ablation in Analytical Chemistry." *Analytical Chemistry* 85, no. 13 (2013): 6162–6177.
27. Geyer, T. J., and W. A. Weimer. "Parametric Effects on Plasma Emission Produced during Excimer Laser Ablation of $YBa_2Cu_3O_7x$." *Applied Spectroscopy* 44, no. 10 (1990): 1659–1664.
28. Lynds, L., B. R. Weinberger, D. M. Potrepka, G. G. Peterson, and M. P. Lindsay. "High-Temperature Superconducting Thin Films: The Physics of Pulsed Laser Ablation." *Physica C: Superconductivity* 159, no. 1–2 (1989): 61–69.
29. Karl, H., and B. Stritzker. "Reflection High-Energy Electron Diffraction Oscillations Modulated by Laser-Pulse Deposited $YBa_2Cu_3O_7$–x." *Physical Review Letters* 69, no. 20 (1992): 2939–2942.
30. Lippmaa, M., N. Nakagawa, M. Kawasaki, S. Ohashi, Y. Inaguma, M. Itoh, and H. Koinuma. "Step-Flow Growth of $SrTiO_3$ Thin Films with a Dielectric Constant Exceeding 10^4." *Applied Physics Letters* 74, no. 23 (1999): 3543–3545.
31. Angusmacleod, H. "Recent Developments in Deposition Techniques for Optical Thin Films and Coatings." In *Optical Thin Films and Coatings*, 3–25. Elsevier, 2013.
32. Morosanu, C., V. Dumitru, E. Cimpoiasu, and C. Nenu. "Comparison between DC and RF Magnetron Sputtered Aluminum Nitride Films." In *Diamond Based Composites*, 127–132. Springer, 1997.
33. Dumitru, V., C. Morosanu, V. Sandu, and A. Stoica. "Optical and Structural Differences between RF and DC AlxNy Magnetron Sputtered Films." *Thin Solid Films* 359, no. 1 (2000): 17–20.
34. Livage, J., C. Sanchez, M. Henry, and S. Doeuff. "The Chemistry of the Sol-Gel Process." *Solid State Ionics* 32–33 (1989): 633–638.
35. Livage, J., and D. Ganguli. "Sol–Gel Electrochromic Coatings and Devices: A Review." *Solar Energy Materials and Solar Cells* 68, no. 3–4 (2001): 365–381.
36. Bokov, D., A. T. Jalil, S. Chupradit, W. Suksatan, M. J. Ansari, I. H. Shewael, G. H. Valiev, and E. Kianfar. "Nanomaterial by Sol-gel Method: Synthesis and Application." *Advances in Materials Science and Engineering* (2021): 5102014. https://doi.org/10.1155/2021/5102014

37. Mackenzie, J. D. "Applications of the Sol-Gel Process." *Journal of Non-Crystalline Solids* 100, no. 1–3 (1988): 162–168.
38. Palve, A. B., C. K. Kasar, and R. S. Khadayate. "ZnO Thin Films by Sol-Gel Method: A Review." *Journal of Emerging Technologies and Innovative Research* 8, no. 9 (2021): 758–763.
39. Bahuguna, G., N. K. Mishra, P. Chaudhary, A. Kumar, and R. Singh. "Thin Film Coating through Sol-Gel Technique." *Research Journal of Chemical Sciences* 6, no. 7 (2016): 65–72.
40. Mane, R. S., and C. D. Lokhande. "Chemical Deposition Method for Metal Chalcogenide Thin Films." *Materials Chemistry and Physics* 65, no. 1 (2000): 1–31.
41. Nanda, K. K., S. N. Sarangi, and S. N. Sahu. "Visible Light Emission from CdS Nanocrystals." *Journal of Physics D: Applied Physics* 32, no. 17 (1999): 2306–2310.
42. Nair, P. K., and M. T. S. Nair. "Chemically Deposited ZnS Thin Films: Application as Substrate for Chemically Deposited Bi_2S_3, Cu_xS and PbS Thin Films." *Semiconductor Science and Technology* 7, no. 2 (1992): 239–244.
43. Dhumure, S., and C. Lokhande. "Studies on the Preparation and Characterization of Chemically Deposited Ag_2S Films from an Acidic Bath." *Thin Solid Films* 240, no. 1–2 (1994): 1–6.
44. Sartale, S. D., and C. D. Lokhande. "Preparation and Characterization of As_2S_3 Thin Films Deposited Using Successive Ionic Layer Adsorption and Reaction (SILAR) Method." *Materials Research Bulletin* 35, no. 8 (2000): 1345–1353.
45. Pramanik, P., P. K. Basu, and S. Biswas. "Preparation and Characterization of Chemically Deposited Tin(II) Sulphide Thin Films." *Thin Solid Films* 150, no. 2–3 (1987): 269–276.
46. Hariskos, D., M. Ruckh, U. Rühle, T. Walter, H. Schock, J. Hedstrom, and L. Stolt. "A Novel Cadmium-Free Buffer Layer for Cu(In,Ga)Se_2-Based Solar Cells." *Solar Energy Materials and Solar Cells* 41–42 (1996): 345–353.
47. Nair, P. K., V. M. Garcia, A. B. Hernandez, and M. T. S. Nair. "Photoaccelerated Chemical Deposition of PbS Thin Films: Novel Applications in Decorative Coatings and Imaging Techniques." *Journal of Physics D: Applied Physics* 24, no. 8 (1991): 1466–1472.
48. Bhat, P. K., S. R. Das, D. K. Pandya, and K. L. Chopra. "Back Illuminated High Efficiency Thin Film Cu_2S/CdS Solar Cells." *Solar Energy Materials* 1, no. 3–4 (1979): 215–219.
49. Pramanik, P., and S. Biswas. "Deposition of Nickel Chalcogenide Thin Films by Solution Growth Techniques." *Journal of Solid State Chemistry* 65, no. 1 (1986): 145–147.
50. Irani, R., S. M. Rozati, and S. Beke. "Structural and Optical Properties of Nanostructural V_2O_5 Thin Films Deposited by Spray Pyrolysis Technique: Effect of the Substrate Temperature." *Materials Chemistry and Physics* 139, no. 2–3 (2013): 489–493.
51. Salehabadi, A., M. Enhessari, M. I. Ahmad, N. Ismail, and B. D. Gupta. "Fabrication of Sensors." In *Metal Chalcogenide Biosensors: Fundamentals and Applications*, 143–174. Woodhead Publishing Series in Electronic and Optical Materials. Elsevier Science, 2023.
52. Anasthasiya, A. N. A., K. Gowtham, R. Shruthi, R. Pandeeswari, and B. G. Jeyaprakash. "Investigation on V_2O_5 Thin Films Prepared by Spray Pyrolysis Technique." *International Journal of Nanoscience* 17, no. 3 (2018): 1760037.
53. Srimathi, S. N., and S. M. Mayanna. "Electroplating of Thin Films of Magnetic Fe-Ni Alloys." *Materials Chemistry and Physics* 11, no. 4 (1984): 351–364.
54. Yurgens, A. *Lab Demo: Electroplating*. FKA195: Nanoscale Science & Technology – Thin Films & Materials. Chalmers University of Technology, November 2004. https://fy.chalmers.se/~yurgens/FKA196/lab%20exercises/PM_plating.pdf
55. Carlsson, J.-O., and P. M. Martin. "Chemical Vapor Deposition." In *Handbook of Deposition Technologies for Films and Coatings*, 314–363. Elsevier, 2010.

56. Eftekhari, A., P. Jafarkhani, and F. Moztarzadeh. "High-Yield Synthesis of Carbon Nanotubes Using a Water-Soluble Catalyst Support in Catalytic Chemical Vapor Deposition." *Carbon* 44, no. 7 (2006): 1343–1345.
57. Tavares, J., E. J. Swanson, and S. Coulombe. "Plasma Synthesis of Coated Metal Nanoparticles with Surface Properties Tailored for Dispersion." *Plasma Processes and Polymers* 5, no. 8 (2008): 759–769.
58. Qin, C., and S. Coulombe. "Organic Layer-Coated Metal Nanoparticles Prepared by a Combined Arc Evaporation/Condensation and Plasma Polymerization Process." *Plasma Sources Science and Technology* 16, no. 2 (2007): 240–249.
59. Ahmad, N. "Application of Thin Films in High-Voltage and Power Electronics." 2025.
60. Jilani, A., M. S. Abdel-wahab, and A. H. Hammad. "Advance Deposition Techniques for Thin Film and Coating." In *Modern Technologies for Creating the Thin-Film Systems and Coatings.* IntechOpen, 2017.

3 Advances in Synthesis, Mechanical Properties, and Corrosion Resistance of Metallic Glass Coatings

Ashutosh Sahu, Venkata Sushma Chinta, Nishant Kumar, and Lavish Kumar Singh

3.1 INTRODUCTION

Bulk metallic glasses (BMGs) or glassy alloys are advanced materials with substantially higher tensile/compressive strength, greater hardness and wear resistance, and higher corrosion resistance than conventional alloys due to the absence of long-range atomic periodicity [1–3]. These materials are synthesized either via rapid solidification processing (RSP) techniques such as melt spinning, copper mould casting, suction casting, pressure die casting, and so on or by consolidating amorphous powders in powder metallurgy route [1–3]. However, a majority of glass-forming configurations necessitate cooling rates to be varied from 104 k/s to 106 k/s via RSP route, and it is very difficult to maintain such high cooling rates throughout the entire cross section during quenching. Thus, dimensions are generally restricted in micrometre to millimetre range in RSP route [1]. Many researchers have tried to synthesize bulk amorphous alloys by consolidating gas atomized or mechanically alloyed amorphous powders or milled amorphous ribbons via powder metallurgy route [1, 4]. Still, it has been observed that the dimensions are limited to a maximum of 20 mm diameter and 5–10 mm thickness [1, 4]. In addition, this class of materials fails catastrophically after crossing the elastic limit under externally applied load due to rapid propagation of shear bands [1–3, 5–7]. Thus, metallic glasses usually do not have any practical industrial applications due to size limitation and a lack of plasticity although these materials exhibit attractive mechanical properties. On the other hand, coatings of metallic glasses on metallic substrates can be practically utilized in many industrial applications attributed to their higher hardness, elastic modulus, tribological characteristics, corrosion resistance, and adhesiveness [8–14]. Various researchers have reported about superior hardness, tribological properties, and corrosion resistance of the glassy alloy coatings [8–14]. For example, 1.2 mm $Fe_{57}Co_{8}Ni_{8}Zr_{10}Si_{4}B_{13}$ glassy coating was made on AISI 1045 alloy via laser cladding, which exhibited hardness of 1270 HV [9]. Similarly, $Zr_{59.3}Cu_{28.8}Al_{10.4}Nb_{1.5}$ glassy coating produced on AA2011-T6 aluminium alloy via layer-wise laser additive manufacturing technique exhibited 20

DOI: 10.1201/9781003750659-3

times higher wear resistance than the substrate [10]. $Fe_{73}Cr_2Si_{11}B_{11}C_3$ amorphous alloy coating produced by atmospheric plasma spraying exhibited substantially improved resistance to corrosion as compared to the low carbon steel [11].

Coatings of metallic glasses exhibit attractive mechanical and physical properties, and an overall idea about various commonly used techniques to produce glassy coatings along with their advantages and drawbacks and the development of mechanical and physical properties with respect to the evolution of different phases during coating would be helpful for people working in this field. Hence, the primary objective of this chapter is to deliver an overall idea of the techniques to produce glassy coatings and the assessment of mechanical characteristics and corrosion resistance. Metallic glass coatings can be produced by different techniques, among which laser cladding, plasma spraying, high-velocity oxy-fuel, and cold spraying are widely used [4–6]. These techniques are discussed with their advantages and drawbacks in the following sections, and a comparative study has also been provided. Among different properties of metallic glass coatings, mechanical properties and corrosion resistance are very important because the coating has to sustain wear and tear and also should not degrade in the corrosive environment. Thus, effort has been made to discuss the mechanical of glassy coatings. The flow chart shown in Figure 3.1 illustrates the topics covered in this chapter.

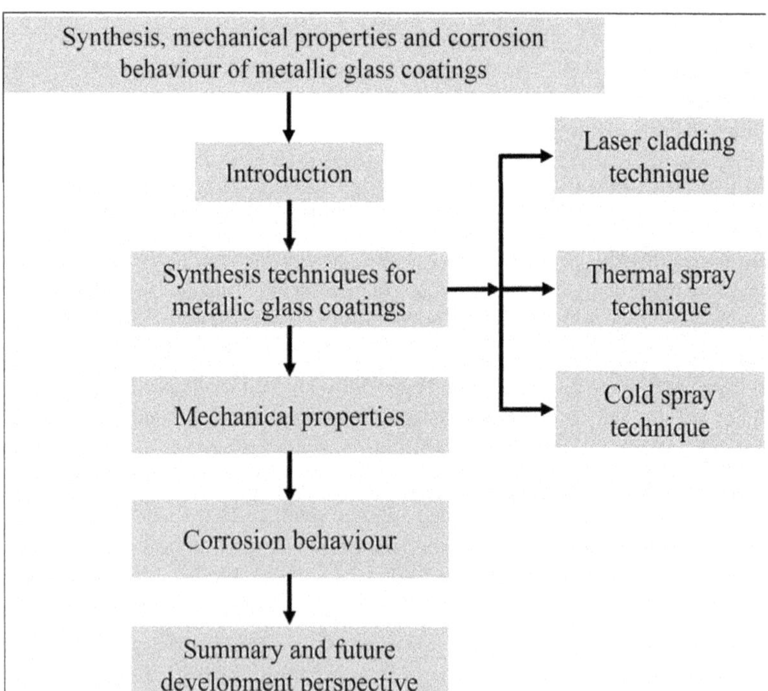

FIGURE 3.1 Flow chart describing the contents in the chapter.

3.2 SYNTHESIS TECHNIQUES

Coatings of amorphous alloys can be made using different techniques such as laser processing, thermal spray techniques such as high-velocity oxy-fuel (HVOF), cold spraying, and so on. All these techniques have their own advantages and drawbacks in terms of coating thickness, adhesion strength, cost-effectiveness, and formation of different phases. Among various techniques laser cladding, plasma spraying, high-velocity oxy-fuel, and cold spraying are very commonly used. Thus, all these techniques are discussed in detail along with their advantages and limitations in the subsequent sections.

3.2.1 Laser Cladding Technique

Laser cladding works on the principle of melting powder layers on the substrate by a focused beam of photons [15, 16]. A laser beam melts the powder coating material along with a small layer of substrate to produce a coating of 50 μm–2 mm thick [17]. The molten material rapidly cools to form a coating on the substrate. Mainly, two techniques are used in laser cladding process viz. (i) two-step process and (ii) one-step process [15–17]. In the two-step route, some pre-alloy powder or a powder mixture is first placed on the substrate surface, and a laser beam scans the powder bed. The powder starts melting by the heat provided by the beam, and when higher heat is provided a thin stratum of the substrate also melts. Thus, the coating material develops a notable metallurgical bonding with the substrate material. As soon as the laser beam is removed, heat is conducted rapidly through the substrate, and the non-equilibrium phase (amorphous phase) is formed [15, 17]. On the other hand, in the latter process, an additive substance in the form of powder or wire or paste is fed along with the laser beam, which melts the entering material, and a thin layer of the substrate and a clad track is formed [15, 17]. When the powder strikes the melt pool it melts completely, while the powder particles that strike the substrate get adhered and increase the surface roughness. Thus, the parameters viz. flow rate of the feed material and carrier gas, the angle of the nozzle, and so on have to be controlled to minimize the surface roughness and attain good metallurgical bonding between the deposited material and the substrate.

The base material is least affected in the course of the laser cladding procedure due to the rapid heating and cooling induced during laser processing. Figure 3.2 shows the laser cladding set-up of both coaxial and preplaced powder systems [18, 19]. The final thickness and microstructure of deposition depends on several factors such as the effect of material properties on the absorption and geometry of melt pool and parameters of laser used in the cladding process [15–17]. Laser cladding process has several advantages such as homogeneous structures can be obtained, complex shapes can be obtained in a single step, parts are of near net shape and only little final finishing operation is required, parts exhibit good mechanical properties, less time is required for making the prototype and tooling by using the CAD model, and initial characteristics of the base material are least effected. Laser cladding also has some drawbacks such as poor reproducibility attributed to high sensitivity of the methodology to laser power, beam velocity, and powder feed rate. An open loop laser cladding

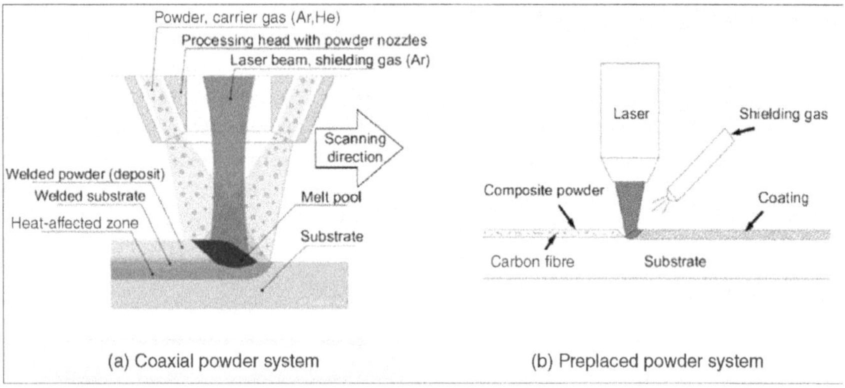

FIGURE 3.2 Laser cladding system for (a) coaxial and (b) preplaced powder system. Figure reproduced with permission from [18, 19].

process with optimal parameters may not produce high-quality cladding because of sporadic or random system disruptions, high investment costs, low efficiency, and inadequate process control.

Several researchers have produced glassy alloy coatings of thickness in the range of 500 μm to 1.5 mm on metallic substrates using laser cladding technique. For example, Wong et al. have coated $AlSi_8Cu_1Mg$ substrate with Ni-Cr-B-Si amorphous alloy of thickness about 0.5 mm using laser cladding technique [8]. Wu et al. have coated AISI 1045 substrate with Fe57Co8Ni8Zr10Si4B13 partially amorphous alloy of maximum thickness 1.2 mm by using laser cladding technique [9]. Sohrabi et al. have produced $Zr_{59.3}Cu_{28.8}Al_{10.4}Nb_{1.5}$ amorphous coating of thickness 300 μm on AA2011-T6 aluminum alloy by using this method [10]. Similarly, Herrera et al. have produced $Fe_{73}Cr_2Si_{11}B_{11}C_3$ coating of thickness ~800 μm on 304 stainless steel substrates [20]. The thickness of the coating depends on the laser power, powder material, and the type of substrate on which the coating is done.

3.2.2 Thermal Spray Technique

In thermal spraying, a heat source melts the amorphous powder material to be deposited on the substrate, and the molten particles from the liquid jet impinge on the substrate, where the molten units rapidly cool resulting in a solid splat of desirable thickness [21, 22]. Figure 3.3 represents a typical set-up for thermal spray process. When completely or partially molten particles collide with the surface of the substrate, they form splats and get adhered to the substrate surface under the impact of molten particles along with rapid quenching. Out of various thermal spray methods, plasma spraying, high-velocity oxy-fuel spraying, and cold spraying techniques are widely used to make glassy coatings on metallic surfaces, and these techniques are discussed in detail in the subsequent sections.

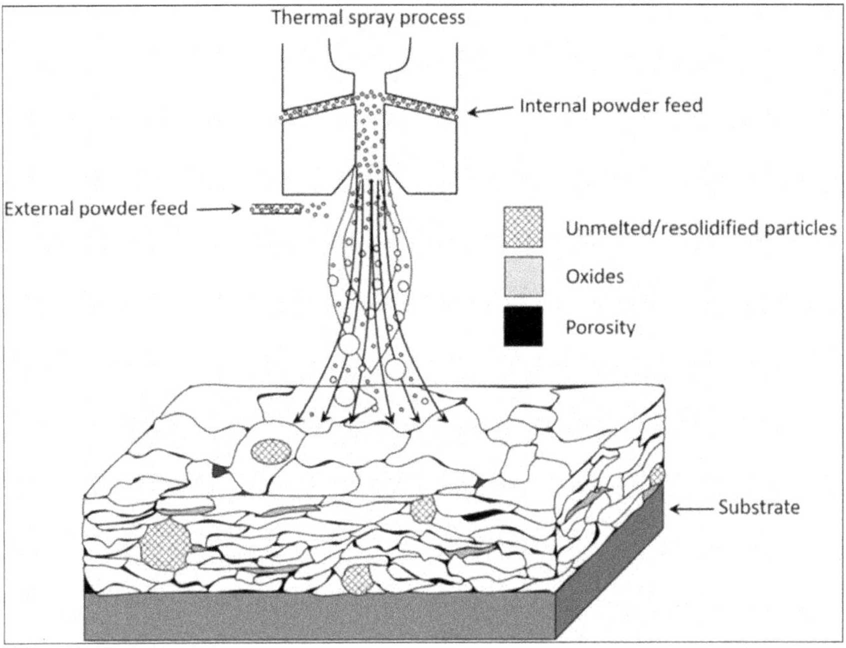

FIGURE 3.3 Schematic showing a set-up for thermal spray process. Open access [21].

3.2.3 Plasma Spraying

In plasma spraying process, direct current or radio-frequency discharge is used to produce plasma [22–26]. The temperature rises up to 8000 K, and glassy power particles are injected within the plasma jet where they are melted and speeded in the direction of the substrate. The powder particles get set down on the substrate surface to form lamellae or splats, and the coating is built by various layers of such splats [22–26]. Figure 3.4 represents the schematic of plasma spray process [23]. In plasma spraying process, two main types of gases are used viz. (i) primary gas such as Ar or N_2 to stabilize arc inside the nozzle and (ii) secondary gas to increase heat conductivity of plasma [23]. The process factors that influence the microstructure and thickness of the coating are the electrical power supply, the flow rate of plasma gases, feed rate of powders, and its size distribution and relative speed of the plasma torch with respect to the substrate [23].

Optimization of these parameters is necessary to get the desired thickness and microstructure to exploit the mechanical properties of the coating. This process has several advantages viz. (i) it can be applied to a wide range of materials, (ii) it has high deposition rates, (iii) adhesive strength is very good with the substrate surface, (iv) it can coat materials of high melting temperatures, and (v) the effect of heat on the substrate is minimum. On the other hand, this process has some drawbacks such

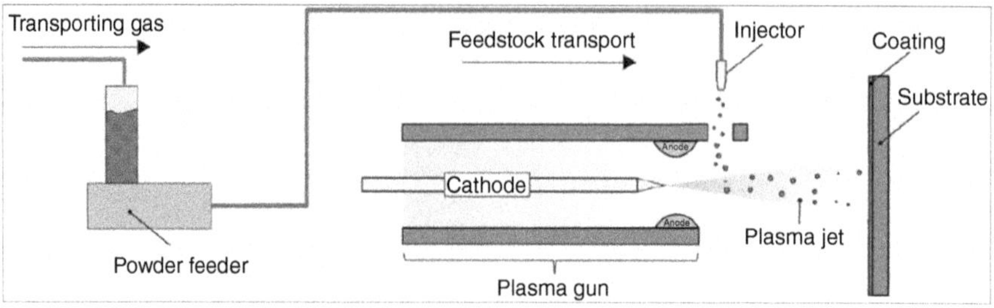

FIGURE 3.4 Schematic showing the set-up for plasma spray process. Open access [23].

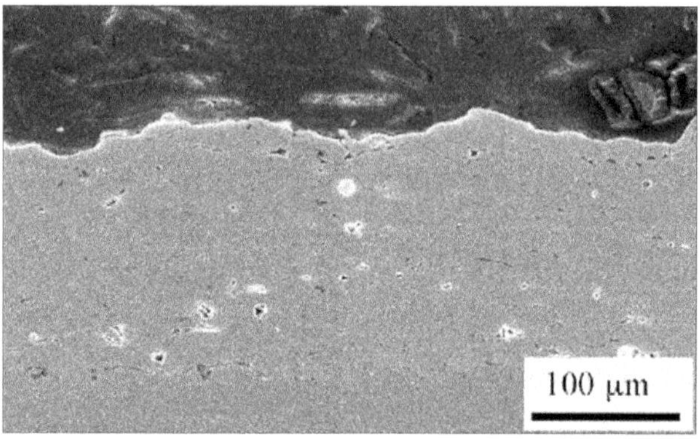

FIGURE 3.5 SEM micrograph showing the $Fe_{50.26}Si_{2.41}B_{2.62}Cr_{23.86}Mo_{20.85}$ metallic glass coating 304 stainless steel [12].

as (i) cost of the equipment is very high; (ii) only external surfaces can be coated, and it is difficult to coat internal surfaces; and (iii) the coating may contain high porosity.

Several researchers have produced metallic glass coatings by using the plasma spray coating technique. For instance, Kobayashi et al. produced $Fe_{50.26}Si_{2.41}B_{2.62}Cr_{23.86}Mo_{20.85}$ glassy coating of thickness 180 μm on 304 stainless steel substrates by using plasma spray technique [12]. The SEM image in Figure 3.5 reveals the glassy coating on steel substrate. Bijalwan et al. have produced $Fe_{73}Cr_2Si_{11}B_{11}C_3$ amorphous coating of thickness 70–170 μm on mild steel substrate using plasma spray technique [11]. An et al. have produced FeWCrNiMoBSiC coating of 180–220 μm thickness on stainless steel [27]. Similarly, Yugeswaran et al. have produced Fe-, Zr- and Ni-based metallic glass coatings of thickness ~200 μm by plasma spraying [28]. Amorphous coatings of thickness varying from 50 μm to 250 μm can be produced by plasma spraying.

3.2.4 High-Velocity Oxy-fuel Spraying

In high-velocity oxy-fuel (HVOF) spraying technique, both fuel gas and oxygen are mixed at a high pressure inside a combustion chamber. This gas mixture is then directed towards the substrate through a small nozzle [23, 29, 30]. The gas

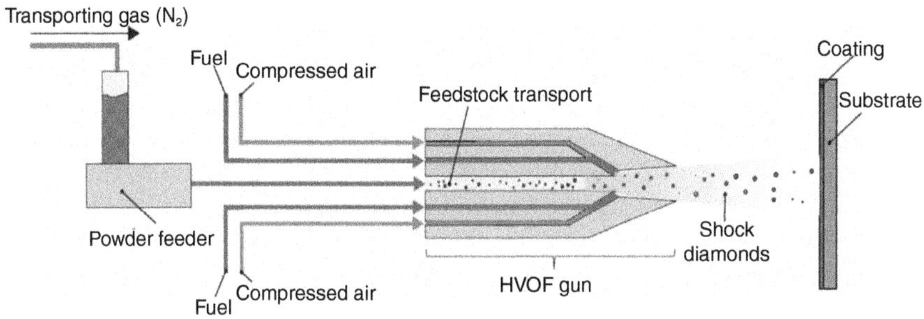

FIGURE 3.6 Schematic showing the set-up for high-velocity oxy-fuel spraying process. Open access [23].

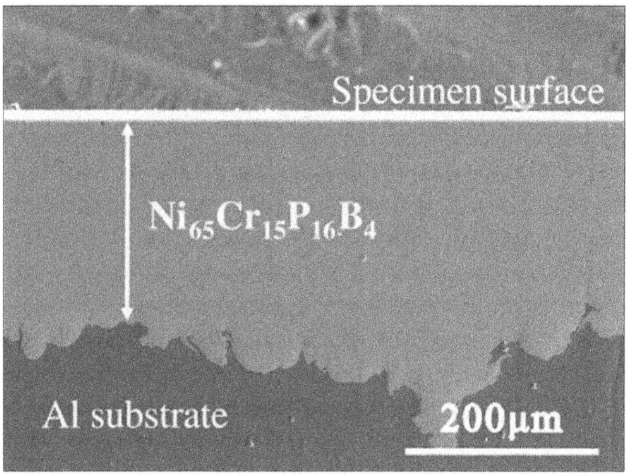

FIGURE 3.7 SEM micrograph revealing the formation of $Ni_{65}Cr_{15}P_{16}B_4$ glassy coating on Al substrate via the HVOF process. Figure reproduced with permission from [30].

attains a velocity of near about 1,000 m/s, and the combustion temperature attained is about 3000 K. The HVOF process attains higher particle velocity and lower particle temperature compared to plasma spraying process. Figure 3.6 shows the schematic of the HVOF set-up. The coating quality is influenced by the distance of spray, feed rate of powder, flow rates of oxygen, fuel, and carrier gas. The coating produced by the HVOF process usually exhibits superior bond strength and less porosity compared to plasma spraying [23, 29, 30]. Many researchers have used the HVOF technique to produce metallic glass coating on metallic substrates [13, 31, 32]. For example, Koga et al. have produced an Fe-Cr-Mo-based glassy coating on a mild steel substrate by the HVOF process [31]. Kim et al. have produced an $Ni_{65}Cr_{15}P_{16}B_4$ glassy coating of thickness around 200 μm on an Al substrate by the HVOF process, as shown in Figure 3.7 [32]. Similarly, Zhang et al. have produced Fe-10W-4Cr-3Ni-2Mo-4B-4Si-1C amorphous coatings on 1Cr18Ni9Ti stainless steel cylinders [13].

3.3 COLD SPRAY TECHNIQUE

Cold spray is a novel technique used to form coatings on metallic surfaces. This technique has several advantages such as prevention of residual stresses, oxidation, and unwanted chemical reactions compared to thermal spray coating techniques, because it utilizes kinetic energy for deposition instead of thermal energy [33–35]. This method uses a convergent-divergent nozzle to accelerate solid powder particles faster than sonic velocity. Particles impinge the substrate at speeds varying from 300 m/s to 1200 m/s. A schematic diagram of the cold spray process is displayed in Figure 3.8.

The ultimate temperature and velocity of sprayed particles are determined by the nozzle geometry and powder qualities, which have an impact on the coating's microstructure, mechanical, and physical attributes. Cold spray is classified as a solid-state deposition processing technology since the gas stream's temperature stays below the melting point of the particles [33–35]. High-pressure cold spray (HPCS) and low-pressure cold spray (LPCS) are the two primary types of cold spray methodology [33–35]. Powder particles in LPCS are injected with a low-pressure gas from the diverging portion of the nozzle, whereas particles in HPCS are injected with a high-pressure gas from the throat of the spray nozzle.

Cold spray process has many advantages such as (i) high amount of amorphous phase can be preserved because this process doesn't involve heating of the powder materials or the substrate, (ii) distortion of parts during the process is minimum, (iii) the bonding strength of particles with the substrate is good and comparable with the previous processes, (iv) the coatings produced via cold spray exhibit low porosity and good corrosion resistance. This process has also some limitations such as spraying complex shapes is difficult. Several researchers have produced metallic glass coatings on different metallic surfaces. For example, List et al. have produced $Fe_{44}Co_6Cr_{15}Mo_{14}C_{15}B_6$ amorphous coating using the cold spray technique [36]. Similarly, Henao et al. have produced $Al_{88}Ni_6Y_{4.5}Co_1La_{0.5}$ glassy coating of thickness around 400 μm on aluminium-7075-T6 alloy substrate by utilizing the cold spray technique [37]. Su et al. have produced $Fe_{48}Cr_{15}Mo_{14}C_{15}B_6Y_2$ amorphous coating of thickness 900 μm on alloy steel plate by cold spraying [38].

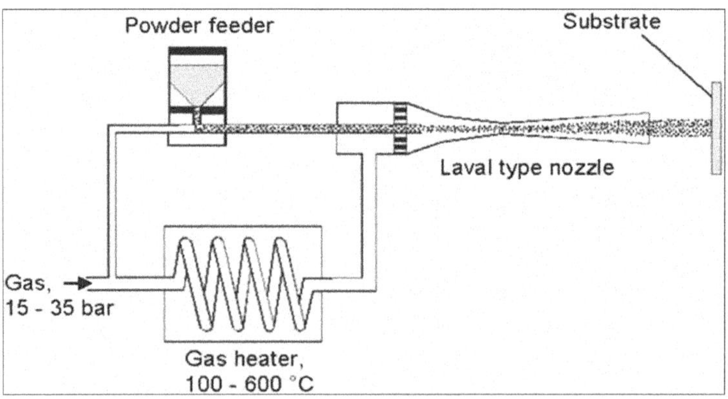

FIGURE 3.8 Schematic representing the cold spray processing technique. Figure reproduced with permission from [33].

3.4 MECHANICAL PROPERTIES

Metallic glasses exhibit around two to three times higher tensile/compression strength, higher hardness, and lower elastic modulus in comparison to the commonly available crystalline materials, attributed to a lack of long-range atomic order [1–7]. Thus, the glassy alloy coatings exhibiting superior hardness, wear resistance, and tensile strength are quite attractive for various industrial applications. Mechanical properties of amorphous coatings are studied at both micro/nano and macro levels. The micro/nano mechanical properties are studied by using Vickers microhardness tester and nanoindentation/nanoscratch tests, whereas, the macro-mechanical properties are studied by using tensile/compression tests etc. Vickers microhardness tests are conducted at low loads (usually 50 gf) so that the indenter won't contact the substrate. Although Vickers microhardness can provide information about the overall hardness of the glassy coatings, it cannot provide information about localized mechanical properties that depend on the localized microstructure. Thus, nanoindentation and nanoscratch tests are conducted to analyse localized mechanical properties. Many researchers have analysed mechanical properties of different metallic glass coatings at macro, micro, and nano levels. For example, Kobayashi et al. have reported Vickers hardness of 1100 HV for $Fe_{50.26}Si_{2.41}B_{2.62}Cr_{23.86}Mo_{20.85}$ glassy coating formed by plasma spraying technique [12]. Similarly, Pandey et al. have studied the hardness of different Fe-based glassy coating by using instrumented microhardness tester and reported hardness values in the range of 7.5 GPa to 9.2 GPa depending on the plasma spray coating parameters [14]. Liu et al. and Pang et al. have studied the nano-mechanical properties of $Zr_{47}Cu_{31}Al_{13}Ni_9$ and $Cu_{49.3}Zr_{50.7}$ glassy coatings respectively via nanoindentation studies [39, 40]. Wear resistance of the glassy coatings can be studied either at the micro level using nanoscratch test or at the macro level using the ball-on-disc test. For example, Zhou et al. have performed nanoscratch test on $Zr_{61}Cu_{17.5}Ni_{10}Al_{7.5}Si_4$ glassy coating. Pandey et al. have studied the wear resistance of Fe-based glassy coating by using the ball-on-disc test [14].

3.5 CORROSION BEHAVIOUR

Metallic glasses exhibit higher corrosion resistance in comparison to the crystalline counterpart, owing to a lack of periodicity in atomic arrangement and consequent absence of grain boundaries, which are known to be the potential locations for commencement of corrosion [1–7]. Usually, coatings are done on metallic substrates to protect them from corrosive environment and, hence, testing of corrosion resistance is very important. Researchers have performed corrosion tests in different electrolytes for different metal-based metallic glasses. For example, corrosion behaviour of Al-based glassy coatings is generally studied in different concentrations of NaCl solution [37, 41]. Lahiri et al. have reported that corrosion resistance of $Al_{90.05}Y_{4.4}Ni_{4.3}Co_{0.9}Sc_{0.35}$ amorphous coating synthesized via cold spraying was found to be higher than the 6061-aluminum substrate in 0.01 N NaCl and 0.1 N NaCl solutions, as shown by the polarization curves depicted in Figure 3.9 [41]. Similarly, Henao et al. found superior corrosion resistance of $Al_{88}Ni_6Y_{4.5}Co_1La_{0.5}$ cold-sprayed amorphous coating as compared to Al-7075-T6 substrate in 0.6 M NaCl solution [37]. Usually, acidic electrolytes of various concentrations are used in case of Zr-,

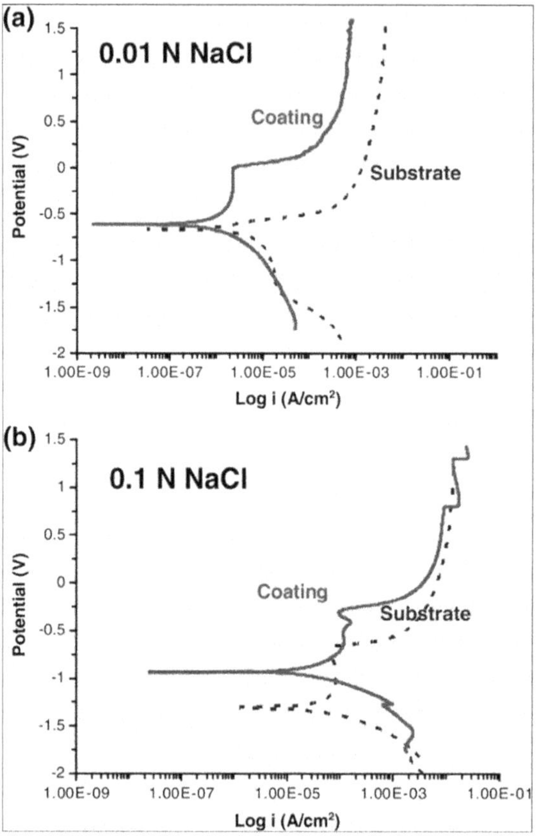

FIGURE 3.9 Potentiodynamic polarization curves of $Al_{90.05}Y_{4.4}Ni_{4.3}Co_{0.9}Sc_{0.35}$ coating formed via cold spray and Al-6061 substrate in (a) 0.01 N NaCl electrolyte and (b) 0.1 N NaCl electrolyte. Figure reproduced with permission from [41].

Ti-, and Fe-based metallic glass coatings on iron steel substrates [42–44]. Ningshen et al. have reported superior corrosion resistance of $Zr_{59}Ti_3Cu_{20}Al_{10}Ni_8$ glassy coating in 1 M, 6 M, and 11.5 M HNO_3 solutions compared to 304L stainless steel substrate [42]. Liu et al. have reported better corrosion resistance of $Fe_{62}Ni_3Cr_4Mo_2W_3Si_6B_{17}C_3$ glassy coating compared to 316L stainless steel substrate in aqueous solution of 3.5% NaCl and 1 N HCl [43]. Similarly, Li et al. studied and found higher corrosion resistance of $Ti_{32.8}Zr_{30.2}Ni_{5.3}Cu_9Be_{22.7}$ glassy alloy coating in comparison to 304L stainless steel substrate; the test being carried out in HNO_3 electrolyte [44].

3.6 SUMMARY AND FUTURE DEVELOPMENT PERSPECTIVE

Metallic glasses exhibit attractive mechanical and physical properties due to the lack of periodicity in the arrangement of atoms. In this perspective, metallic glass coatings can be fruitful in many potential industrial applications. There are techniques such as laser cladding, plasma spraying, thermal spraying, cold spraying, and so on to produce metallic glass coatings. Optimization of parameters is important in case of all these processes to obtain desired coating thickness and to retain the amorphous phase of the coatings, so that the desired mechanical properties can be obtained. Sophisticated instrumentation or artificial intelligence can be applied to obtain better control during the coating process, so that the speed, heat, and material flow during coating can be easily controlled. Metallic glass coatings exhibit better hardness and corrosion resistance as compared to the crystalline substrates. Although it has been reported that glassy coatings display improved corrosion resistance compared to the crystalline substrate, considerable study needs to be carried out to comprehend the corrosion behaviour of coating consisting of both glassy and crystalline phases.

REFERENCES

1. C. Suryanarayana, A. Inoue, *Bulk metallic glasses*, CRC Press, Taylor and Francis Group, Boca Raton (2011) 1–525.
2. J.F. Loffler, Bulk metallic glasses, *Intermetallics*, 11 (2003) 529–540.
3. J.J. Kruzic, Bulk metallic glasses as structural materials: A review, *Advanced Engineering Materials*, 18 (2016) 1308–1331.
4. A. Sahu, R.S. Maurya, T. Laha, Advances in synthesis and characterization of aluminum-based amorphous alloys: A review, *Advanced Engineering Materials*, 26 (2024) 2301150.
5. M.M. Trexler, N.N. Thadhani, Mechanical properties of bulk metallic glasses, *Progress in Materials Science*, 55 (2010) 759–839.
6. J. Eckert, J. Das, S. Pauly, C. Duhamel, Mechanical properties of bulk metallic glasses and composites, *Journal of Materials Research*, 22 (2007) 285–301.
7. D.V.L. Luzgin, L.V.L. Luzgina, A. Yu. Churyumov, Mechanical properties and deformation behavior of bulk metallic glasses, *Metals*, 3 (2013) 1–22.
8. T.T. Wong, G.Y. Liang, B.L. He, C.H. Woo, Wear resistance of laser-clad Ni-Cr-B-Si alloy on aluminium alloy, *Journal of Materials Processing Technology*, 100 (2000) 142–146.
9. X. Wu, Y. Hong, Fe-based thick amorphous-alloy coating by laser cladding, *Surface and Coatings Technology*, 141 (2001) 141–144.
10. N. Sohrabi, R.S. Panikar, J. Jhabvala, A.R. Buch, S. Mischler, R.E. Loge, Laser coating of a Zr-based metallic glass on an aluminum substrate, *Surface & Coatings Technology*, 400 (2020) 126223.
11. P. Bijalwan, C. Singh, A. Kumar, K. Sarkar, N. Rani, T. Laha, A. Banerjee, K. Mondal, Corrosion behaviour of plasma sprayed Fe based metallic glass Fe73Cr2Si11B11C3 (at%) coatings in 3.5% NaCl solution, *Journal of Non-Crystalline Solids*, 567 (2021) 120913.
12. A. Kobayashi, S. Yano, H. Kimura, A. Inoue, Fe-based metallic glass coatings produced by smart plasma spraying process, *Materials Science and Engineering B*, 148 (2008) 110–113.
13. H. Zhang, Y. Hu, G. Hou, Y. An, G. Liu, The effect of high-velocity oxy-fuel spraying parameters on microstructure, corrosion and wear resistance of Fe-based metallic glass coatings, *Journal of Non-Crystalline Solids*, 406 (2014) 37–44.
14. K.K. Pandey, P. Singh, A. Pathak, P. Bijalwan, M. Dutta, A. Banerjee, A.K. Keshri, Tailoring the mechanical, tribological and corrosion behavior of Fe-based metallic glass coating synthesized using atmospheric plasma spraying, *Journal of Thermal Spray Technology*, 31 (2022) 1342–1354.

15. L. Zhu, P. Xue, Q. Lan, G. Meng, Y. Ren, Z. Yang, P. Xu, Z. Liu, Recent research and development status of laser cladding: A review, *Optics & Laser Technology*, 138 (2021) 106915.
16. E. Williams, N. Lavery, Laser processing of bulk metallic glass: A review, *Journal of Materials Processing Technology*, 247 (2017) 73–91.
17. R. Vilar, Laser cladding, *Journal of Laser Applications*, 11 (1999) 64–78.
18. F. Wirth, K. Wegener, A physical modeling and predictive simulation of the laser cladding process, *Additive Manufacturing*, 22 (2018) 307–319.
19. J. Lei, C. Shi, S. Zhou, Z. Gu, L.C. Zhang, Enhanced corrosion and wear resistance properties of carbon fiber reinforced Ni-based composite coating by laser cladding, *Surface & Coatings Technology*, 334 (2018) 274–285.
20. J.E.G. Herrera, J. Henao, D.G.E. Arbelaez, J.M.G. Carmona, C.F. Martinez, R.S. Fernandez, J.C. Castuera, C.A.P. Salas, J.M.A. Orozco, Laser cladding deposition of a Fe-based metallic glass on 304 stainless steel substrates, *Journal of Thermal Spray Technology*, 31 (2022) 968–979.
21. D.T. Martin, M.R. Rad, A. McDonald, T. Hussain, Beyond traditional coatings: A review on thermal-sprayed functional and smart coatings, *Journal of Thermal Spray Technology*, 28 (2019) 598–644.
22. P. Fauchais, Understanding plasma spraying, *Journal of Physics D: Applied Physics*, 37 (2004) R86–R108.
23. L. Latka, L. Pawlowski, M. Winnicki, P. Sokolowski, A. Malachowska, S. Kozerski, Review of functionally graded thermal sprayed coatings, *Applied Sciences*, 10 (2020) 5153.
24. A. Vardelle, C. Moreau, N.J. Themelis, C. Chazelas, A perspective on plasma spray technology, *Plasma Chemistry and Plasma Processing*, 35 (2015) 491–509.
25. H. Herman, Plasma spray deposition processes, *MRS Bulletin*, 13 (1988) 60–67.
26. P. Fauchais, G. Montavon, Plasma spraying: From plasma generation to coating structure, *Advances in Heat Transfer*, 40 (2007) 205–344.
27. Y. An, G. Hou, J. Chen, X. Zhao, G. Liu, H. Zhou, J. Chen, Microstructure and tribological properties of iron-based metallic glass coatings prepared by atmospheric plasma spraying, *Vacuum*, 107 (2014) 132–140.
28. S. Yugeswaran, A. Kobayashi, Metallic glass coatings fabricated by gas tunnel type plasma spraying, *Vacuum*, 110 (2014) 177–182.
29. K. Anusha, B.C. Routara, S. Guha, A review on high-velocity oxy-fuel (HVOF) coating technique, *Journal of the Institution of Engineers (India): Series D*, 104 (2023) 831–848.
30. M. Oksa, E. Turunen, T. Suhonen, T. Varis, S.P. Hannula, Optimization and characterization of high velocity oxy-fuel sprayed coatings: Techniques, materials, and applications, *Coatings*, 1 (2011) 17–52.
31. Y. Koga, K. Sakata, D. Oka, K. Kuwatori, I. Narita, H. Miyahara, Formation of Fe-Cr-Mo alloy metallic glass coating using high-velocity oxy-fuel apparatus with hydrogen gas and its microstructural transition at elevated temperatures, *Materials Transactions*, 58 (2017) 1444–1450.
32. S. Kim, S. Yamaura, Y. Shimizu, K. Nakashima, T. Igarashi, A. Makino, A. Inoue, Production of Ni65Cr15P16B4 metallic glass-coated bipolar plate for fuel cell by high velocity oxy-fuel (HVOF) spray coating method, *Materials Transactions*, 51 (2010) 1609–1613.
33. F. Gartner, T. Stoltenhoff, T. Schmidt, H. Kreye, The cold spray process and its potential for industrial applications, *Journal of Thermal Spray Technology*, 15 (2006) 223–232.
34. A. Srikanth, G. Md. T. Basha, B. Venkateshwarlu, A brief review on cold spray coating process, *Materials Today: Proceedings*, 22, (2020) 1390–1397.

35. P. Khatake, R. Taluja, M.S. Kumar, M.M. Reddy, F.H.A. Ataby, S. Sood, P. Sonia, Cold spray coating: A review of material systems and future perspectives, *Materials Today: Proceedings*. https://doi.org/10.1016/j.matpr.2023.11.015
36. A. List, F. Gartner, T. Schmidt, T. Klassen, Impact conditions for cold spraying of hard metallic glasses, *Journal of Thermal Spray Technology*, 21 (2012) 531–540.
37. J. Henao, A. Concustell, I.G. Cano, S. Dosta, N. Cinca, J.M. Guilemany, T. Suhonen, Novel Al-based metallic glass coatings by cold gas spray, *Materials & Design*, 94 (2016) 253–261.
38. J. Su, J. Kang, W. Yue, G. Mac, Z. Fu, L. Zhu, D. She, H. Wang, C. Wang, Comparison of tribological behavior of Fe-based metallic glass coatings fabricated by cold spraying and high velocity air fuel spraying, *Journal of Non-Crystalline Solids*, 522 (2019) 119582.
39. F.X. Liu, Y.F. Gao, P.K. Liaw, Rate-dependent deformation behavior of Zr-based metallic-glass coatings examined by nanoindentation, *Metallurgical and Materials Transactions A*, 39A (2008) 1862–1867.
40. J.J. Pang, M.J. Tan, K.M. Liew, C. Shearwood, Nanoindentation study of size effect and loading rate effect on mechanical properties of a thin film metallic glass Cu49.3Zr50.7, *Physica B*, 407 (2012) 340–346.
41. D. Lahiri, P.K. Gill, S. Scudino, C. Zhang, V. Singh, J. Karthikeyan, N. Munroe, S. Seal, A. Agarwal, Cold sprayed aluminum based glassy coating: Synthesis, wear and corrosion properties, *Surface and Coatings Technology*, 232 (2013) 33–40.
42. S. Ningshen, U.K. Mudali, R. Krishnan, B. Raj, Corrosion behavior of Zr-based metallic glass coating on type 304L stainless steel by pulsed laser deposition method, *Surface & Coatings Technology*, 205 (2011) 3961–3966.
43. G. Liu, Y. An, Z. Guo, J. Chen, G. Hou, J. Chen, Structure and corrosion behavior of iron-based metallic glass coatings prepared by LPPS, *Applied Surface Science*, 258 (2012) 5380–5386.
44. C. Li, D. Chen, W. Chen, L. Wang, D. Luo, Corrosion behavior of TiZrNiCuBe metallic glass coatings synthesized by electro spark deposition, *Corrosion Science*, 84 (2014) 96–102.

4 Defects in Thin Films – Part I

Adeeba Mirza and Abhilasha Mishra

4.1 INTRODUCTION

The very thin layers of materials which deposit on different substrates having a thickness between a few nanometres and several micrometres are known as thin films [1]. These coatings are done on a substrate's surface to obtain changes in optical, thermal, electrical, and mechanical properties [2]. Thin films are used in different fields such as in semiconductors, anti-reflective coatings, solar cells, and as protective coatings. Techniques such as chemical vapour deposition (CVD), plasma-enhanced CVD, physical vapour deposition (PVD), sputtering and atomic layer deposition (ALD) are utilized to create thin films [3]. Every individual method suggests inimitable challenges and advantages specific to various material requirements and applications. Besides, the chapter will highlight the various applications of thin films, indicating their serious role in technological developments [4]. From their use in manufacturing high-efficiency photovoltaic cells for solar energy to integrated circuits and semiconductor devices across various industries, thin films are at the forefront of innovation and are also helping to reduce costs [5]. In medical implant devices, corrosion-resistant and wear-resistant coatings are created using thin films to make the surface biocompatible [6]. The advanced properties of thin films rise because of their compact dimensionality, which can easily affect their mechanical, chemical, optical, and electrical behaviours [7]. These films form the building blocks of microchips for smartphones, computers, and other digital devices. In the optic field, the thin films are utilized to develop coatings on mirrors, lenses, and on other optical devices [8]. These coatings can control reflection, light transmission, and absorption, showing applications in reflective coatings for mirrors, in glasses for anti-reflective coatings, and in various layers of display screens and camera lenses [9]. In electronics, thin film coatings (TFCs) are very crucial for the manufacturing of transistors, integrated circuits, semiconductors, and other electrical devices.

Crucial imperfections in thin films known as defects in coatings affect the dependability and performance of these films in different fields of application [10]. They are used broadly in industries because of their capacity to increase surface properties like optical clarity, electrical conductivity, and mechanical durability [11]. However, the existence of defects, such as cracks, pinholes, dislocations, voids, and inclusions, can destabilize these advantages, lead to the failure of the coated products, degradation in functionality, and reduced lifespan [12]. Considering the origins, nature, and consequences of these defects is important for improving fabrication methods and ensuring the performance and high quality of thin-film coatings for

 DOI: 10.1201/9781003750659-4

advancements in technological applications [13]. Defects in thin films arise from a combination of material properties, deposition processes, and environmental factors, significantly impacting the film's structural, mechanical, and functional characteristics [14]. Growth-related defects, like uneven nucleation, coalescence issues, and columnar structures, often lead to non-uniform coverage and void formation. Intrinsic defects, such as vacancies, interstitials, and dislocations, disturb the crystal lattice, while stress-induced defects emerge from thermal expansion mismatches and lattice strain between the thin film and substrate [15]. Film quality is further degraded on introduction of impurities by contamination of oxygen, moisture, or residual gases during deposition. Prevention of proper crystallization is raised by deposition technique issues, like non-uniform material distribution and high deposition rates leading to poor adhesion. High-quality thin film production is managed by post-deposition processes, including improper annealing and oxidation, which can intensify these defects, making essential careful control over each step. The act and reliability of these films across various advanced applications are profoundly influenced by these crucial imperfections that are defects in thin film coatings [16, 17]. Thin films are broadly used in optics, electronics, and materials science because of their capability to knowingly improve the surface properties such as optical clarity, mechanical durability, and electrical conductivity [18]. In optics, defects such as cracks, inclusions, and voids can cause absorption and scattering of light, fading the reflectivity and transparency which are important for devices like mirrors, lenses, and displays [19]. However, the existence of defects extending from cracks and pinholes to voids can severely destabilize these advantages, leading to reduction in functionality, small lifespan, and subsequent failure of the coated materials [20]. These defects can arise during several stages of the fabrication procedure, including environmental conditions, substrate preparation, and deposition methods [21]. Defects like pinholes can arise because of dust particles or contaminants on the surface of the substrate, whereas cracking arises because of thermal stresses at the time of deposition [22]. The nature and influence of these defects may vary depending upon the application part. For instance, irregularities on the surface can be the reason for scattering and can reduce the optical efficiency [23]. In the case of electronic gadgets, even small defects can disturb the circuit pathways. Apparently, understanding the mechanisms behind the formation of defects and then developing the approaches to eliminate or diminish these imperfections are important for advancing the thin film technology [24]. Improved fabrication techniques, innovative materials, and rigorous quality control are all part of current efforts to optimize performance, address these challenges, enhance durability, and ensure the high quality of thin-film coatings in their respective applications [25]. This chapter gives an idea about the types of defects that occur in thin films, their origin, and characterization techniques used to identify these defects and the mitigation strategies to overcome these defects.

4.2 IMPORTANCE OF THIN FILMS IN SCIENCE AND TECHNOLOGY

Thin films are important for several scientific and technological progresses (see Figure 4.1) due to their unmatched properties and functionalities. In electronics

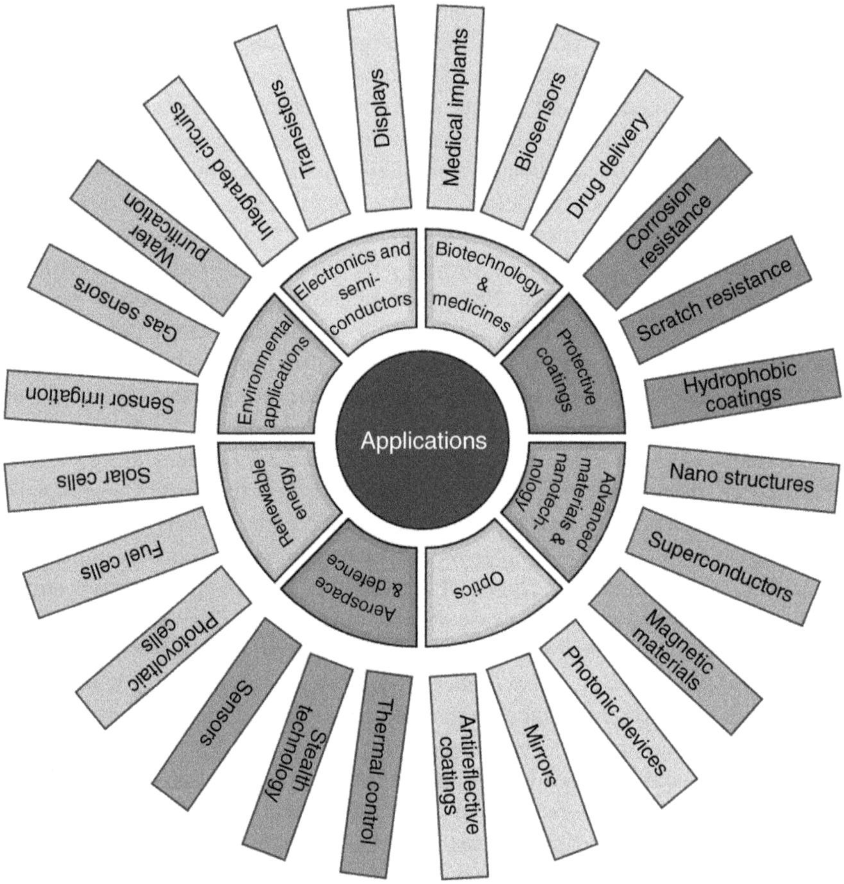

FIGURE 4.1 Applications of thin films.

and semiconductors, they are important in the modification of integrated transistors, circuits, and displays which have high-resolution [26]. Thin films find application in optics through high-reflectivity mirrors, anti-reflective coatings, and photonic devices. Renewable energy technologies influence thin films in solar cells and fuel cells for cost-effective energy solutions [27]. In biotechnology and medicine, thin films enhance medical implants, biosensors, and drug delivery systems. Protective coatings made from thin films provide corrosion and scratch resistance in numerous applications [28]. Environmental technologies utilize thin film membranes for water purification and gas sensors. In nanotechnology and advanced materials, thin film coatings (TFC) contribute to the advanced development of superconductors, magnetic materials, and nanostructures [29]. Aerospace and defence sectors use thin films for thermal control, protection in harsh environments, and stealth technology. Their versatility makes thin films essential in current modern science and technology [30].

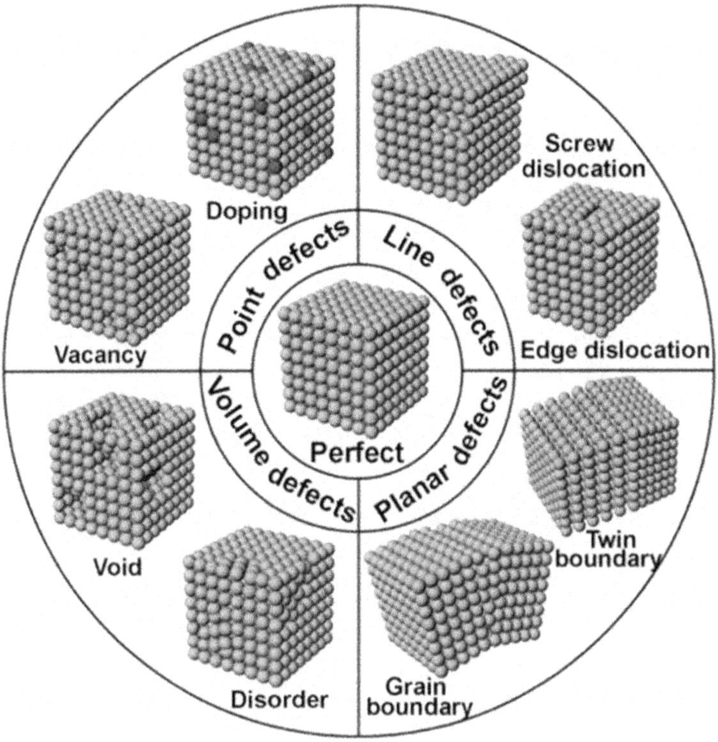

FIGURE 4.2 Types of defects (Reproduced with permission [54]).

4.3 TYPES OF DEFECTS IN THIN FILMS

There are many types of defects as shown in Figure 4.2 and briefly described in the following.

4.3.1 Surface Defects

These defects in thin films refer to imperfections and irregularities that occur on the surface of films, which can affect the entire performance and properties of thin films [31]. The main types of defects in surface are as follows.

i) Roughness

The uneven texture of thin film on the substrate in a micro to nano scale is known as surface roughness. It can arise due to non-uniform deposition processes, intrinsic material properties, or imperfections in substrate [32]. Surface roughness directly affects the optical properties by reducing the transparency, increasing the scattering of light, and changing the reflectivity range [33]. The mechanical properties like friction and wear rates enhance with increasing roughness.

ii) Cracks

The splits and fractures in thin films that extend completely or partially in the thickness of the film are known as cracks [34]. This happens because of thermal stresses, intrinsic tensile strength, and mechanical stress which arise during the handling process and affect the mechanical properties leading to strength reduction and increasing the possibility of delamination [35]. Cracks on surfaces disrupt the electrical continuity required in films for maintaining conductivity.

iii) Pinholes

These are the minor, open pits or holes on the surface that penetrate through the film, characteristically on the order of micrometres to nanometres in diameter which occur due to non-uniform deposition, contamination on the substrate, or incomplete coverage during deposition [36]. Pinholes are the pathways for gases, moisture, and contaminants which reduce the efficiency of protective coatings [37]. Also, they lead to short circuit paths in insulating layers, affecting the electrical properties by compromising the dielectric integrity.

iv) Delamination

The peeling or separation of TFC from the surface of substrate is known as delamination which arises due to poor adhesion. It occurs because of intrinsic stress, surface contamination, and thermal cycling [38]. It affects the mechanical properties due to loss of structural integrity.

4.3.2 Point Defects

The atomic-scale imperfections in thin films that arise inside the crystal lattice of the substrate are known as point defects [39, 40]. These defects can intentionally influence the electrical, physical, chemical, and optical properties of thin films. Here are the main types of point defects.

i) Vacancies

This defect arises when an atom is missing from the crystal lattice; it happens throughout the deposition process due to radiation damage, high-temperature processing, or insufficient atomic supply [41]. When vacancies are developed on the surfaces, they can work as charge carrier's trapping sites and can reduce the mobility of ions, thereby affecting the conductivity [42]. This defect can also affect the mechanical properties by weakening the material's surface as it disturbs the crystal structure.

ii) Interstitials

When extra atoms occupy positions in the crystal lattice, it is termed as interstitials, where they usually do not belong, normally in the spaces between steady atomic sites. It can arise from high-energy methods like rapid quenching or ion implantation through deposition. These can cause internal stress and lattice distortion, affecting the mechanical strength and durability [43]. In electrical properties, these defects can introduce bandgaps, which affect the semiconductor properties.

iii) Substitutional Impurities

These impurities occur when external atoms replace host atoms in the lattice. This defect arises because of intentional doping for modification of material properties or unintentional contamination through the deposition process which affects the electrical properties by severely varying the carrier concentration and conductivity, particularly in semiconductors [44]. Also, they can weaken or strengthen the material, depending on the nature and size of the impurity atoms.

iv) Interstitial Impurities

These impurities arise when foreign atoms occupy interstitial sites in the lattice which frequently result from intentional doping or contamination during film growth. These defects can cause internal stresses and lattice distortion, possibly leading to mechanical failure. They can also create recombination centres or deep-level traps, disturbing the semiconductor's electrical performance [45]. They may enhance the chemical reactivity of the substrate by affecting corrosion resistance and stability.

4.3.3 Defect Clusters

The groups of closely spaced point defects that connect with each other, making more complex structures, are known as defect clusters in thin films. These clusters can severely affect the performance and material properties of TFC [46]. Here is a description of defect clusters, their types, and formation.

i) Vacancy Clusters

When a high concentration of missing atoms arises to create a group of vacancies, it is known as a vacancy cluster. They can form by irradiation or at high temperatures, where vacancies are mobile and can easily coalesce. These clusters create stress fields that lead to weakening the mechanical strength and also forming cracks and voids.

ii) Interstitial Clusters

These are the aggregation of interstitial atoms that form a concentrated region of additional atoms in the lattice [48]. Characteristically, they form under conditions of high energy, for example, high-dose irradiation or ion implantation, where interstitial atoms can migrate and group together. They can cause important lattice distortion that leads to mechanical failure and possible internal stress. This defect can also work as scattering centres and can reduce the carrier mobility in films of semiconductors.

iii) Mixed Clusters

These are the combinations of interstitial sites and vacancies that come together to form more complex defect structures. These often arise through processes that generate both interstitials and vacancies, like high-energy particle bombardment. This defect can lead to complex stress fields, disturbing the durability and mechanical integrity of the film. Also, they influence the thermal conductivity by scattering phonons, which can influence thermal management in devices [47–49].

4.3.4 Interface Defects

Interface defects in thin films are irregularities or imperfections that occur at the boundary between the TFC and the substrate or between different layers in a multilayer structure. These defects can suggestively influence overall properties and performance of the thin film. Here is an explanation of interface defects, their formation, and their effects on the properties of TFC.

i) Dislocations

These are interfacial misalignments in crystal lattice, where the substrate doesn't entirely fit with that of atomic arrangement of the film. These are characterized by variances in thermal expansion coefficients and during the growth of epitaxial layers, arising because of the variance among the substrate and thin film. The electrical properties of interface defects include the trapping of charge carriers through dislocations resulting in mobility reduction and thus affecting the electrical performance of semiconductor devices. Additionally, they can act as points of mechanical weakness, potentially leading to cracking or delamination under stress.

ii) Grain Boundaries

These are areas where different crystal orientations connect, often occurring at the interface between dissimilar layers or between the film and substrate. They occur naturally during the growth of polycrystalline films where grains grow independently and meet at the interface. Electrical conductivity can be lowered when grain boundaries act as barriers to charge carrier movement. These regions are often more chemically reactive, leading to increased susceptibility to corrosion and oxidation.

iii) Voids and Gaps

These are hollow gaps at the interface where the adhesion of the film is not properly achieved, thus resulting in voids and gaps during deposition, contamination, or differences in surface energy between the film and the substrate. Voids can act as thermal insulators, which in turn affect heat transfer across the interface. It may also decrease the adhesion strength causing the film more prone to peeling or delamination.

iv) Chemical Inhomogeneities

These can arise during deposition processes or the successive thermal treatments causing interdiffusion of elements resulting in alterations of chemical composition at the interface. These alterations affect the refractive index and optical transparency. They can initiate localized electronic states concerning the electronic properties in the matrix of the film.

v) Interface Roughness

The roughness of substrate surface and deposition and post deposition process variations and treatments give rise to the uneven boundary involving the substrate and film occurring due to the irregularity of the surface topology at the interface. Optical properties include the increasing surface light, which affects the reflectivity and transparency. Whereas under the mechanical load it experiences failure due to stress concentration [50–52].

4.3.5 Bulk Defects

Bulk defects in thin films refer to imperfections that occur within the interior (basically in the bulk) of the TFC, away from the surface or interfaces. These defects can suggestively affect the physical, electrical, optical, and mechanical properties of the TFC [53]. Here is an explanation of bulk defects, their types, and their effects on different properties of these films (TFC).

i) Voids

These are empty spaces or cavities within thin films arising due to incomplete packing of atoms, gas entrapment, or shrinkage during cooling throughout the deposition process. The cracking and failure of the film occurs due to the reduction in mechanical strength and durability. Suspended continuousness of conductive paths leads to the reduced conductivity and increased electrical resistance. By acting as thermal insulators, thermal conductivity is decreased.

ii) Dislocations

These are misalignments of atoms and line defects within the crystal structure of the TFC. It arises from lattice mismatch, thermal stress, or plastic deformation on or after deformation. It builds stress fields which can reduce mechanical strength and increase the likelihood of fracture, establish localized states affecting light absorption and emission, and behaves as scattering centres for charge carriers, decreasing the carrier mobility and affecting electrical conductivity.

iii) Precipitates

These are the minute, distinct phases or clusters of atoms within the bulk of the film that differ in composition from the surrounding material. They form during deposition or subsequent thermal treatments due to supersaturation of solute atoms. It can either strengthen or weaken the film, depending on the nature and distribution of the precipitates. Furthermore, they can act as scattering centres or recombination sites, affecting the performance of semiconductor films. Also, they affect the optical transparency and refractive index.

iv) Interstitials

These are the extra atoms situated at non-lattice sites within crystal structure. These are developed from high-energy processes, for instance, ion implantation or rapid quenching during deposition. Due to lattice distortion and internal stress, mechanical failure occurs. The introduction of localized states within the bandgap affects electrical performance. Diffusion rates are enhanced due to the increasing chemical reactivity.

v) Vacancies

Empty lattice sites are created because of the absent atoms in crystal lattice occurring due to thermal fluctuations, irradiation, or non-stoichiometric deposition processes. They act as blockers for charge carriers lowering mobility and conductivity and create fragile points in the lattice reducing the mechanical strength [55]. Increasing atomic mobility affects the stability and performance of the TFC.

4.3.6 Nano-Structural Defects

Nano-structural defects in thin films are imperfections that occur at the nanoscale, influencing the material's properties at very small dimensions. These defects can have momentous impacts on the electrical, physical, optical, and mechanical property of TFC. Here is an explanation of nano-structural defects, their types, and their effects on TFC properties.

i) Nano-Voids

These are minute, hollow spaces existing in the film within nanometre scale, formed during the deposition process caused by the incomplete packing of atoms, gas entrapment, or during post-deposition treatments. They make the film vulnerable, that is more prone to cracking and failure resulting in lowered mechanical strength. Additionally, they may function as heat insulators which lower thermal conductivity. Electrical resistance can be increased by disrupting the continuousness of conductive films [56].

ii) Nano Cracks

These are minute cracks existing in the film but for a minimum width in nanometres, resulting from thermal stress, mechanical stress, or intrinsic stress during deposition. They create vulnerable points that lead to crack propagation under stress. This compromises the ability of the film to act as a barrier against moisture, gases, and chemicals.

iii) Nanoprecipitates

These are unique phase or composition of nanoscale clusters that exist within the thin film. They are formed as a result of supersaturation of solute atoms occurring throughout deposition or thermal treatments. Precipitates have the possibility to either strengthen or weaken the film, depending on their make-up and distribution [57]. The impact on the performance of semiconductor films is characterized by the effect on charge carrier movements. Also, it may introduce scattering centres, which in turn affect the transparency and refractive index.

iv) Grain Boundaries

Grain boundaries between nanometer-sized grains in polycrystalline films constitute a major source of defect formation during the film growth process. They affect the hardness and brittleness of the film. Furthermore, they can act as barriers to charge carrier movement, enhance diffusion along the boundaries lowering electrical conductivity and affecting material stability respectively.

v) Surface Roughness

Surface roughness refers to a film's surface irregularities in the nanoscale range resulting from non-uniform deposition, substrate imperfections, or intrinsic material properties [58]. The intensifying light scattering reduces transparency and alters reflectivity. It can affect wear resistance by influencing friction. Moreover, it affects the adhesion of films to other substrate layers.

4.3.7 Process-Induced Defects

Process-induced defects in thin films are imperfections that arise during the manufacturing or processing stages of thin film deposition [59]. These defects can severely impact the properties and performance of TFC. Here is an explanation of process-induced defects, their types, and their effects on TFC properties.

i) Inclusions

These are foreign particles or impurities in thin films generated during the process of deposition or successive steps of processing. It can result from the contamination of deposition sources, substrate surfaces, or processing environments. Inclusions can act as stress concentrators and scatter light, leading to mechanical weakness, potential failure, and decreasing optical transparency, thereby affecting the optical performance. Electrical conductivity and device performance can be affected by introducing charge traps or scattering centres.

ii) Crystalline Defects

These are imperfections including dislocations, stacking faults, and grain boundaries that exist inside the film's crystal lattice structure. Reduced mechanical strength and increased susceptibility to failure can result from deposition processes with non-uniform growth conditions, cooling rates, or lattice mismatches. Crystalline defects make the film fragile leading to reduced mechanical strength and increased susceptibility to fracture [60]. Additionally, these defects can introduce absorption or emission features, altering the optical characteristics of the film. Dislocations and grain boundaries can disturb the mobility of charge carrier and conductivity in semiconductor films.

iii) Surface Contamination

These defects may result from inadequate cleaning of deposition equipment, exposure to ambient air, or contamination during handling. They occur when foreign materials are adsorbed or deposited onto the surface of the thin film during processing. Surface contamination can impair adhesion between the thin film and the substrate or subsequent layers, leading to delamination or peeling. The surface chemistry, affecting reactivity, stability, and performance, is altered by contaminants.

iv) Cracks and Delamination

These are fractures or separations within the thin film or between layers in a multilayer structure. Thermal stresses, mechanical stresses during deposition, or inadequate adhesion between layers result in cracks and delamination. They weaken the film compromising mechanical integrity, leading to failure under stress. Also, they initiate pathways for moisture, gases, and chemicals, reducing the effectiveness of protective coatings [61].

4.4 PROPERTIES AFFECTED BY DEFECTS IN THIN FILMS

Defects in TFC are irregularities or imperfections that can occur throughout the deposition process or because of continuous environmental exposure [62, 63]. These

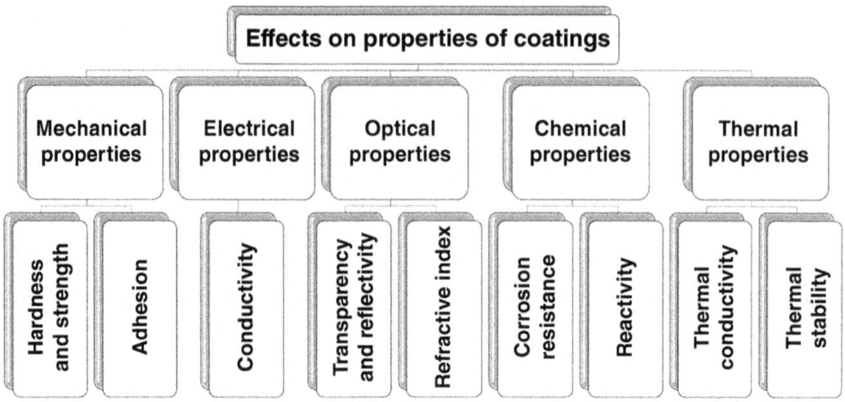

FIGURE 4.3 Different properties of coatings affected by defects.

defects can significantly affect the performance and properties of thin-film layers. The properties of the coatings are also influenced by these defects [64]. A range of material properties may be influenced by defects such as cracks and grain boundaries, as illustrated in Figure 4.3.

4.4.1 Mechanical Properties

Defects influence the mechanical strength, hardness, and durability of thin films, affecting their structural integrity and resistance to mechanical deformation, wear, and fracture [65]. Defects like voids and dislocations can weaken the film, decreasing its mechanical strength and hardness. Voids and cracks at the interface may lead to poor adhesion of the TFC on to the substrate's surface and cause delamination [66]. Grain boundaries, dislocations, and voids act as stress concentrators, leading to localized mechanical failures such as cracking, delamination, and film buckling [67].

4.4.2 Electrical Properties

The electrical conductivity, resistivity, and career mobility of material are affected by defects including vacancies, interstitials, grain boundaries, and dislocations which introduce localized states within the material's bandgap. The electrical conductivity of the TFC is either degraded or improved by impurities and point defects depending on the nature of the defect [14, 68]. The leakage currents, electrical resistance, and reduced device performance in electronic and optoelectronic applications are enhanced, as these defects can work as trapping sites for charge carriers [69].

4.4.3 Optical Properties

These properties of TFC affect light absorption, emission, and transmission by the alteration of defects [70]. For example, the material's optical absorption spectrum, photoluminescence efficiency, and light-emitting properties can be changed by defects like vacancies and impurity atoms which create energy levels within the bandgap [71]. The transparency and reflectivity of optical coatings are decreased and varied from scattering of light caused by grain boundaries and surface roughness [72]. In optical coatings and photonic devices, defects can lead to unsuitable optical losses, reduced efficiency, and degraded spectral performance [2].

4.4.4 Chemical Properties

The chemical stability and corrosion resistance is impacted by the defects in thin films by altering their surface reactivity, chemical bonding, and susceptibility to environmental degradation [73]. Defects in surfaces include vacancies and dangling bonds that can assist as active sites for reactions, leading to increased surface oxidation, corrosion, and ruin over time [74]. In protective type of coatings and sensors applications, where high chemical stability is required, defects can compromise the thin film's continuing performance and reliability [75, 76].

4.4.5 Thermal Properties

Dislocations and cracks can lead to instability and thermal stresses, which trigger the film to degrade under thermal cycling [77]. The thermal conductivity of the TFC is reduced by the scattering of photons by defects like voids.

4.4.6 Device Performance

Defects directly influence the performance and functionality of thin film-based devices, including transistors, solar cells, sensors, and memory devices [78]. In electronic devices, defects affect charge transport, carrier mobility, and switching characteristics, leading to variations in device behaviour, power consumption, and signal-to-noise ratio [79]. Similarly, in photovoltaic devices, defects impact carrier recombination, open-circuit voltage, and fill factor, affecting device efficiency and output power.

4.4.7 Process Yield and Reliability

Defects play a crucial role in determining the process yield and reliability of TFC fabrication processes [80]. High-defect densities can lead to increased manufacturing costs, lower production yields, and reduced device performance consistency [81]. Defect-related failures and reliability issues can result in product recalls, warranty claims, and customer dissatisfaction, impacting the overall success and competitiveness of TFC-based technologies in the market. These defects significantly affect the functionality of thin films across different applications including electrical, optical, mechanical, and chemical properties, as well as device performance, process yield, and reliability [16]. Table 4.1 lists the different materials properties affected by different types of defects in TFC. Understanding and mitigating the impact of defects

TABLE 4.1
Properties affected by different types of defects in thin films

S. No.	Substrate	Technique	Findings	Properties Affected	References
1.	Glass	Pulse laser deposition (PLD)	Crystal defect	Optical property	[82]
2.	Silicon	Dipping technique	Lattice defect	Optical property	[83]
3.	Glass	PLD	High crystallinity, substrate-dependent electrical properties	Electrical, optical	[84]
4.	ZnO:Al thin films	PLD	Improved crystallinity, lower stress, enhanced optical properties	Optical, structural	[85]
5.	Soda-lime, quartz	Sol-gel	Sodium ion impact, crystallization, surface morphology	Structural, morphological	[86]
6.	Glass	Sol-gel	Oxygen-related defects, electronic transport	Electrical	[87]
7.	Glass	SILAR (sequential ionic layer adsorption, and reaction)	High crystallinity, low resistivity with ethanol	Electrical, optical	[88]
8.	$LaAlO_3$ (LAO) substrate	Ion beam sputtering	Oxygen vacancies, surface roughness, strain effects	Structural, morphological	[89]
9.	Glass	RF sputtering	Lowest resistivity at RT, bandgap changes, surface morphology	Electrical, optical	[90]
10.	Glass, Si	RF sputtering	Crystallinity improvement with temperature, optical properties	Structural, optical	[91]
11.	Glass	ALD	Point defects, electrical and optical properties	Electrical, optical	[92]
12.	ZnSnO films	PLD	Oxygen vacancies affect electrical properties, TFT performance	Electrical, optical	[93]
13.	MgO monolayers	DFT	Bandgap reduction due to oxygen vacancies	Electronic, optical	[94]
14.	Oxide heterostructure	Various	3D oxygen vacancy distribution, multiferroic properties	Structural, magnetic	[95]
15.	$Hf_{0.5}Zr_{0.5}O_{2-x}$ HZO thin films	PLD	Oxygen vacancies affect ferroelectricity	Ferroelectric, electrical	[96]
16.	$La_{0.7}Sr_{0.3}MnO_3$ thin films	PLD	Conductive paths, memristive switching	Electrical, magnetic	[97]

TABLE 4.1 (*Continued*)
Properties affected by different types of defects in thin films

S. No.	Substrate	Technique	Findings	Properties Affected	References
17.	$SrFeO_3$-δ thin film	Sputtering	Lattice parameter changes, dielectric properties	Structural, dielectric	[98]
18.	$La_{0.7}Ca_{0.3}MnO_{3-x}$ thin film	Pulsed laser deposition (PLD)	Metal-insulator transition temperature shift	Electrical, structural	[99]
19.	$LaNiO_3$-x(LNO) film grown on $SrTiO_3$ (STO)	PLD	Electronic phase separation, surface transformation	Electrical, structural	[100]
20.	NiO_x film	Sputtering	Oxygen vacancy formation, work function changes	Electrical, chemical	[101]

is essential for optimizing thin film materials and technologies for diverse industrial and technological applications.

4.5 CONCLUSION AND FUTURE PROSPECTS

The study of defects in thin films is essential for understanding their properties, performance, and reliability across various applications. Throughout this chapter, we have explored the different types of defects that can happen in thin films, including vacancies, dislocations, grain boundaries, and surface imperfections. These defects can appear during thin film deposition, processing, or post-treatment stages, impacting the material's electrical, optical, mechanical, and chemical properties. To develop mitigation strategies which are aimed at reducing defect density and improving thin film quality, understanding the formation mechanisms and characteristics of defects is critical. Researchers and engineers can minimize defect formation and enhance thin film performance by implementing optimized deposition parameters, substrate cleaning techniques, surface passivation methods, and post-deposition annealing treatments.

REFERENCES

1. Liang, S., Schwartzkopf, M., Roth, S. V., and Müller-Buschbaum, P. State of the art of ultra-thin gold layers: Formation fundamentals and applications. *Nanoscale Advances* 4.12 (2022): 2533–2560.
2. Liu, B., Wang, C., Bazri, S., Badruddin, I. A., Orooji, Y., Saeidi, S., and Mahian, O. Optical properties and thermal stability evaluation of solar absorbers enhanced by nano-structured selective coating films. *Powder Technology* 377 (2021): 939–957.
3. Jones, A. C., and Hitchman, M. L. (Eds.). *Chemical vapour deposition: Precursors, processes and applications*. Royal Society of Chemistry, 2009.

4. Gausemeier, J., Echterhoff, N., Kokoschka, M., and Wall, M. *Thinking ahead the future of additive manufacturing*. Future Applications, 2011.
5. Chopra, K. L., Paulson, P. D., and Dutta, V. Thin-film solar cells: An overview. *Progress in Photovoltaics: Research and Applications* 12.2–3 (2004): 69–92.
6. Sathish, M., Radhika, N., and Saleh, B. Current status, challenges, and future prospects of thin film coating techniques and coating structures. *Journal of Bio- and Tribo-Corrosion* 9.2 (2023): 35.
7. Pacchioni, G. Two-dimensional oxides: Multifunctional materials for advanced technologies. *Chemistry–A European Journal* 18.33 (2012): 10144–10158.
8. Ştefanov, T., Maraka, H. V. R., Meagher, P., Rice, J., Sillekens, W., and Browne, D. J. Thin film metallic glass broad-spectrum mirror coatings for space telescope applications. *Journal of Non-Crystalline Solids: X* 7 (2020): 100050.
9. Raut, H. K., Ganesh, V. A., Nair, A. S., and Ramakrishna, S. Anti-reflective coatings: A critical, in-depth review. *Energy & Environmental Science* 4.10 (2011): 3779–3804.
10. Ramachandran, A. M., Singh, M., Thampi, A. S., and Asok, A. Antireflective coatings and optical filters. In *Metal oxides for next-generation optoelectronic photonic and photovoltaic applications*. Elsevier, 2024: 343–371.
11. Devasahayam, S., and Hussain, C. M. Thin-film nanocomposite devices for renewable energy current status and challenges. *Sustainable Materials and Technologies* 26 (2020): e00233.
12. Sastri, V. S. *Challenges in corrosion: Costs, causes, consequences, and control*. John Wiley & Sons, 2015.
13. Sabzehmeidani, M. M., Gafari, S., and Kazemzad, M. Concepts, fabrication and applications of MOF thin films in optoelectronics: A review. *Applied Materials Today* 38 (2024): 102153.
14. Park, J. S., Kim, S., Xie, Z., and Walsh, A. Point defect engineering in thin-film solar cells. *Nature Reviews Materials* 3.7 (2018): 194–210.
15. Buchheit, T., Takashima, K., Spolenak, R., Minor, A., and Rd, O. C. In *Thin films-stresses and mechanical properties XI*. Materials Research Society Symposium Proceedings Series, 2005: 875.
16. Napari, M., Huq, T. N., Hoye, R. L., and MacManus-Driscoll, J. L. Nickel oxide thin films grown by chemical deposition techniques: Potential and challenges in next-generation rigid and flexible device applications. *InfoMat* 3.5 (2021): 536–576.
17. Meda, U. S., Bhat, N., Pandey, A., Subramanya, K. N., and Raj, M. L. A. Challenges associated with hydrogen storage systems due to the hydrogen embrittlement of high strength steels. *International Journal of Hydrogen Energy* 48.47 (2023): 17894–17913.
18. Ukoba, K., and Jen, T. C. *Thin films, atomic layer deposition, and 3D Printing: Demystifying the concepts and their relevance in industry 4.0*. CRC Press, 2023.
19. Ristau, D., and Ehlers, H. Thin film optical coatings. In *Springer handbook of lasers and optics*. New York: Springer, 2007: 373–396.
20. Hintze-Bruening, H., and Leroux, F. Nanocomposite based multifunctional coatings. In *New advances in vehicular technology and automotive engineering*. IntechOpen, 2012: 55.
21. Oviroh, P. O., Akbarzadeh, R., Pan, D., Coetzee, R. A. M., and Jen, T. C. New development of atomic layer deposition: Processes, methods and applications. *Science and Technology of Advanced Materials* 20.1 (2019): 465–496.
22. Panjan, P., Drnovšek, A., Gselman, P., Čekada, M., and Panjan, M. Review of growth defects in thin films prepared by PVD techniques. *Coatings* 10.5 (2020): 447.
23. Berginski, M., Hüpkes, J., Schulte, M., Schöpe, G., Stiebig, H., Rech, B., and Wuttig, M. The effect of front ZnO: Al surface texture and optical transparency on efficient light trapping in silicon thin-film solar cells. *Journal of Applied Physics* 101.7 (2007).

24. Min, J., Choi, Y., Kim, D., and Park, T. Beyond imperfections: Exploring defects for breakthroughs in perovskite solar cell research. *Advanced Energy Materials* 14.6 (2024): 2302659.
25. Tofail, S. A., Koumoulos, E. P., Bandyopadhyay, A., Bose, S., O'Donoghue, L., and Charitidis, C. Additive manufacturing: Scientific and technological challenges, market uptake and opportunities. *Materials Today* 21.1 (2018): 22–37.
26. Chu, Y., Qian, C., Chahal, P., and Cao, C. Printed diodes: Materials processing, fabrication, and applications. *Advanced Science* 6.6 (2019): 1801653.
27. Ganjoo, A., McCamy, J., Polcyn, A., Ma, Z., and Medwick, P. A. Glass and coatings on glass for solar applications. In *Springer handbook of glass*. Cham: Springer International Publishing, 2019: 1635–1676.
28. Yousaf, S., Alhnan, M. A., Abdallah, A., Abdallah, B., Khan, I., and Ahmed, W. Nanocoatings in medicine: Antiquity and modern times. In *Emerging nanotechnologies for manufacturing*. William Andrew Publishing, 2015: 418–443.
29. Song, Y., and Zhang, W. *Inorganic and organic thin films*. Wiley-VCH, 2021.
30. Barshilia, H. C. Surface modification technologies for aerospace and engineering applications: Current trends, challenges and future prospects. *Transactions of the Indian National Academy of Engineering* 6.2 (2021): 173–188.
31. Gritsenko, K. P., and Krasovsky, A. M. Thin-film deposition of polymers by vacuum degradation. *Chemical Reviews* 103.9 (2003): 3607–3650.
32. Mylvaganam, K., Chen, Y., Liu, W., Liu, M., and Zhang, L. Hard thin films: Applications and challenges. *Anti-abrasive Nanocoatings* (2015): 543–567.
33. Flory, F., Escoubas, L., and Berginc, G. Optical properties of nanostructured materials: A review. *Journal of Nanophotonics* 5.1 (2011): 052502–052502.
34. Lee, Y., Tan, H. H., Jagadish, C., and Karuturi, S. K. Controlled cracking for large-area thin film exfoliation: Working principles, status, and prospects. *ACS Applied Electronic Materials* 3.1 (2020): 145–162.
35. Alkhazali, A., Hamasha, M. M., Khaled, H., Alkhazaleh, A., and Etier, M. Electro-mechanical properties of molybdenum thin film on polyethylene terephthalate subjected to tensile stress. *IEEE Transactions on Device and Materials Reliability* 24.2 (2024): 233–240.
36. Cazzaniga, A. C. Fabrication of thin film CZTS solar cells with Pulsed Laser Deposition. Technical University of Denmark, 2016. https://orbit.dtu.dk/en/publications/fabrication-of-thin-film-czts-solar-cells-with-pulsed-laser-depos/
37. Licari, J. J. Coating materials for electronic applications: Polymers, processing, reliability, testing. Noyes Publications, 2003.
38. Khanna, V. K. Adhesion–delamination phenomena at the surfaces and interfaces in microelectronics and MEMS structures and packaged devices. *Journal of Physics D: Applied Physics* 44.3 (2010): 034004.
39. de Castro Saiki, L. E., and Gomes, G. F. Understanding and mitigating delamination in composite materials: A comprehensive review. *Mechanics of Advanced Materials and Structures* (2024): 1–21.
40. Tuller, H. L., and Bishop, S. R. Point defects in oxides: Tailoring materials through defect engineering. *Annual Review of Materials Research* 41 (2011): 369–398.
41. Morosanu, C. E. *Thin films by chemical vapour deposition*. Vol. 7. Elsevier, 2016.
42. Li, G., Blake, G. R., and Palstra, T. T. Vacancies in functional materials for clean energy storage and harvesting: The perfect imperfection. *Chemical Society Reviews* 46.6 (2017): 1693–1706.
43. Crawford, J. H., and Slifkin, L. M. *Point defects in solids: General and ionic crystals*. Springer Science & Business Media, 2013.
44. Rautray, T. R., Narayanan, R., and Kim, K. H. Ion implantation of titanium based biomaterials. *Progress in Materials Science* 56.8 (2011): 1137–1177.

45. Holt, D. B., and Yacobi, B. G. *Extended defects in semiconductors: Electronic properties, device effects and structures*. Cambridge University Press, 2007.
46. Ahmed, I. *Defects in semiconductors*. The University of Manchester (United Kingdom), 1985.
47. Rudolph, P. Fundamentals and engineering of defects. *Progress in Crystal Growth and Characterization of Materials* 62.2 (2016): 89–110.
48. Zhang, Y., Zhao, S., Weber, W. J., Nordlund, K., Granberg, F., and Djurabekova, F. Atomic-level heterogeneity and defect dynamics in concentrated solid-solution alloys. *Current Opinion in Solid State and Materials Science* 21.5 (2017): 221–237.
49. Wu, Z., Luo, Z., Shen, Y., Zhao, W., Wang, W., Nan, H., and Ni, Z. defects as a factor limiting carrier mobility in WSe2: A spectroscopic investigation. *Nano Research* 9 (2016): 3622–3631.
50. Huang, G., and Mei, Y. Thinning and shaping solid films into functional and integrative nanomembranes. *Advanced Materials* 24.19 (2012): 2517–2546.
51. Gao, Z., Leng, C., Zhao, H., Wei, X., Shi, H., and Xiao, Z. The electrical behaviors of grain boundaries in polycrystalline optoelectronic materials. *Advanced Materials* 36.4 (2024): 2304855.
52. Rana, M. P. S., Singh, F., Joshi, K., Negi, S., and Ramola, R. C. Influence of electronic excitations on structural, optical and electrical properties of undoped and antimony doped tin oxide thin films. *Thin Solid Films* 616 (2016): 34–42.
53. Freund, L. B., and Suresh, S. *Thin film materials: Stress, defect formation and surface evolution*. Cambridge University Press, 2004.
54. Bai, S., Zhang, N., Gao, C., and Xiong, Y. defect engineering in photocatalytic materials. *Nano Energy* 53 (2018): 296–336.
55. Greene, J. E. Thin film nucleation, growth, and microstructural evolution: An atomic scale view. In *Handbook of deposition technologies for films and coatings*. William Andrew Publishing, 2010: 554–620.
56. Kholmanov, I. N., Stoller, M. D., Edgeworth, J., Lee, W. H., Li, H., Lee, J., and Ruoff, R. S. Nanostructured hybrid transparent conductive films with antibacterial properties. *ACS Nano* 6.6 (2012): 5157–5163.
57. Zink, J., Wyrobnik, T., Prinz, T., and Schmid, M. Physical, chemical and biochemical modifications of protein-based films and coatings: An extensive review. *International Journal of Molecular Sciences* 17.9 (2016): 1376.
58. Benetti, G., Banfi, F., Cavaliere, E., and Gavioli, L. Mechanical properties of nanoporous metallic ultrathin films: A paradigmatic case. *Nanomaterials* 11.11 (2021): 3116.
59. Qin, L., Wang, K., Li, X., Zhou, S., and Yang, G. Review of the formation mechanisms and control methods of geometrical defects in laser deposition manufacturing. *Chinese Journal of Mechanical Engineering: Additive Manufacturing Frontiers* 1.4 (2022): 100052.
60. Alexopoulos, P. S., and O'Sullivan, T. C. Mechanical properties of thin films. *Annual Review of Materials Science* 20.1 (1990): 391–420.
61. Tator, K. B., and Lanterman, R. Coating deterioration – a mechanistic overview. *NACE CORROSION* (2016): NACE-2016-7065.
62. Zidarič, T., and Maver, U. Biopolymer thin films as "smart" materials in biomedical applications. *Functional Biomaterials: Design and Development for Biotechnology, Pharmacology, and Biomedicine* 1 (2023): 239–268.
63. Schlom, D. G., Chen, L. Q., Pan, X., Schmehl, A., and Zurbuchen, M. A. A thin film approach to engineering functionality into oxides. *Journal of the American Ceramic Society* 91.8 (2008): 2429–2454.
64. Shirke, N., Ghase, V., and Jamdar, V. Recent advances in stealth coating. *Polymer Bulletin* (2024): 1–30.
65. Lionello, D. F., Ramallo, J. I., Soler-Illia, G. J., and Fuertes, M. C. Mechanical properties of ordered mesoporous oxides thin films. *Journal of Sol-Gel Science and Technology* (2022): 1–26.

66. Garg, R., Gonuguntla, S., Sk, S., Iqbal, M. S., Dada, A. O., Pal, U., and Ahmadipour, M. Sputtering thin films: Materials, applications, challenges and future directions. *Advances in Colloid and Interface Science* (2024): 103203.
67. Zhou, Q., Ren, Y., Du, Y., Hua, D., and Han, W. Cracking and toughening mechanisms in nanoscale metallic multilayer films: A brief review. *Applied Sciences* 8.10 (2018): 1821.
68. Pastor, E., Sachs, M., Selim, S., Durrant, J. R., Bakulin, A. A., and Walsh, A. Electronic defects in metal oxide photocatalysts. *Nature Reviews Materials* 7.7 (2022): 503–521.
69. Li, W., Shi, J., Zhang, K. H., and MacManus-Driscoll, J. L. defects in complex oxide thin films for electronics and energy applications: Challenges and opportunities. *Materials Horizons* 7.11 (2020): 2832–2859.
70. Ezike, S. C. Effect of concentration variation on optical and structural properties of TiO2 thin films. *Journal of Modern Materials* 7.1 (2020): 1–6.
71. Zhang, C., and Lin, J. Defect-related luminescent materials: Synthesis, emission properties and applications. *Chemical Society Reviews* 41.23 (2012): 7938–7961.
72. Hruška, P., More-Chevalier, J., Novotný, M., Čížek, J., Melikhova, O., Fekete, L., and Fitl, P. Effect of roughness and nanoporosity on optical properties of black and reflective Al films prepared by magnetron sputtering. *Journal of Alloys and Compounds* 872 (2021): 159744.
73. Pedanekar, R. S., Shaikh, S. K., and Rajpure, K. Y. Thin film photocatalysis for environmental remediation: A status review. *Current Applied Physics* 20.8 (2020): 931–952.
74. Li, Q., Zhou, Q., Shi, L., Chen, Q., and Wang, J. Recent advances in oxidation and degradation mechanisms of ultrathin 2D materials under ambient conditions and their passivation strategies. *Journal of Materials Chemistry A* 7.9 (2019): 4291–4312.
75. Li, M., Igbari, F., Wang, Z. K., and Liao, L. S. Indoor thin-film photovoltaics: Progress and challenges. *Advanced Energy Materials* 10.28 (2020): 2000641.
76. Abdeen, D. H., El Hachach, M., Koc, M., and Atieh, M. A. A review on the corrosion behaviour of nanocoatings on metallic substrates. *Materials* 12.2 (2019): 210.
77. Putz, B., Wurster, S., Edwards, T. E. J., Völker, B., Milassin, G., Többens, D. M., and Cordill, M. J. Mechanical and optical degradation of flexible optical solar reflectors during simulated low earth orbit thermal cycling. *Acta Astronautica* 175 (2020): 277–289.
78. Ge, S., Sang, D., Zou, L., Yao, Y., Zhou, C., Fu, H., and Wang, C. A review on the progress of optoelectronic devices based on TiO2 thin films and nanomaterials. *Nanomaterials* 13.7 (2023): 1141.
79. Niu, X. *Growth and properties of nanocrystalline germanium and germanium carbide films and photovoltaic devices*. Iowa State University, 2005.
80. Wen, Y., Liu, Y., Guo, Y., Yu, G., and Hu, W. Experimental techniques for the fabrication and characterization of organic thin films for field-effect transistors. *Chemical Reviews* 111.5 (2011): 3358–3406.
81. Han, G., Zhang, S., Boix, P. P., Wong, L. H., Sun, L., and Lien, S. Y. Towards high efficiency thin film solar cells. *Progress in Materials Science* 87 (2017): 246–291.
82. Anjum, A., Ahmed, R., Umar, Z. A., Azzam, S., Hussain, T., Sarwar, M. N., and Baig, M. A. Structure and defects-related optical properties of highly (002)-oriented zinc oxide thin films. *Physica B: Condensed Matter* 644 (2022): 414195.
83. Stroescu, H., Nicolescu, M., Mitrea, D., Tenea, E., Atkinson, I., Anastasescu, M., and Gartner, M. Effect of Al incorporation on the structural and optical properties of sol–gel AZO thin films. *Materials* 16.9 (2023): 3329.
84. Fouad, K., Temam, E. G., Adel, T., and Abd Errahmane, B. Effect of substrates on the structural and optical properties of cobalt-doped ZnO thin films. *Studies in Engineering and Exact Sciences* 5.3 (2024): e13132.
85. Kumar, P., Dharmaprakash, S. M., Kote, M. A., Sandeep, K. M., Patil, P. S., and Bhajantri, R. F. Target-to-substrate distance influenced linear and nonlinear optical properties of a-plane oriented ZnO:Al thin films. *Journal of Materials Science: Materials in Electronics* 32.17 (2021): 22214–22231.

86. Kment, Š., Gregora, I., Kmentová, H., Novotná, P., Hubička, Z., Krýsa, J., and Hrabovský, M. Raman spectroscopy of dip-coated and spin-coated sol–gel TiO2 thin films on different types of glass substrate. *Journal of Sol-Gel Science and Technology* 63 (2012): 294–306.
87. Scalvi, L. V. D. A., Geraldo, V., Messias, F. R., Li, M. S., Santilli, C. V., and Pulcinelli, S. H. Contribution of oxygen related defects to the electronic transport in SnO2 sol-gel films. *Radiation Effects and Defects in Solids* 156.1–4 (2001): 145–149.
88. Yergaliuly, G., Soltabayev, B., Kalybekkyzy, S., Bakenov, Z., and Mentbayeva, A. Effect of thickness and reaction media on properties of ZnO thin films by SILAR. *Scientific Reports* 12.1 (2022): 851.
89. Panomsuwan, G., and Saito, N. Effect of oxygen partial pressure on crystal structure, oxygen vacancy, and surface morphology of epitaxial SrTiO3 thin films grown by ion beam sputter deposition. *Oxygen* 1.1 (2021): 62–72.
90. Waykar, R., Amit, P., Kulkarni, R., Jadhavar, A., Funde, A., Waman, V., and Jadkar, S. Low substrate temperature deposition of transparent and conducting ZnO:Al thin films by RF magnetron sputtering. *Journal of Semiconductors* 37.4 (2016): 043001.
91. Ben Mbarek, I., Chaabouni, F., Selmi, M., Abaab, M., and Rezig, B. Effect of the substrate temperature on the properties of the RF sputtered TiO2 thin films. *Physica Status Solidi C* 7.9 (2010): 2311–2315.
92. Tomasz, A., Łuka, G., Wachnicki, Ł., Rafałjakieła, B. W., Guziewicz, E., Godlewski, M., and Tallarida, G. Optical and electrical characterization of defects in zinc oxide thin films grown by atomic layer deposition. *Optica Applicata* 39.4 (2009).
93. Zhang, J., Lu, J., Lu, Y., Yue, S., Lu, R., Li, X., and Ye, Z. dependence of device behaviours on oxygen vacancies in ZnSnO thin-film transistors. *Applied Physics A* 125 (2019): 1–8.
94. Hazarika, R., and Kalita, B. Effect of oxygen vacancy defects on electronic and optical properties of MgO monolayers: First principles study. *Materials Science and Engineering: B* 286 (2022): 115974.
95. Hunnestad, K. A., Das, H., Hatzoglou, C., Holtz, M., Brooks, C. M., van Helvoort, A. T., and Meier, D. 3D oxygen vacancy distribution and defect-property relations in an oxide heterostructure. *Nature Communications* 15.1 (2024): 5400.
96. Liu, S., Li, X., Li, G., Yan, S., Zhu, Y., Wu, Y., and Tang, M. In situ modulation of oxygen vacancy concentration in Hf0.5Zr0.5O2−x thin films and the mechanism of its impact on ferroelectricity. *Coatings* 14.9 (2024): 1121.
97. Liang, Z. W., Wu, P., Wang, L. C., Shen, B. G., and Wang, Z. H. Conductive path and local oxygen-vacancy dynamics: Case study of crosshatched oxides. *Chinese Physics B* 32.4 (2023): 047303.
98. Enriquez, E., Chen, A., Harrell, Z., Dowden, P., Koskelo, N., Roback, J., and Jia, Q. Oxygen vacancy-tuned physical properties in perovskite thin films with multiple B-site valance states. *Scientific Reports* 7.1 (2017): 46184.
99. Salas-Colera, E., Muñoz-Noval, A., Sebastiani-Tofano, E., Castro, G. R., and Rubio-Zuazo, J. Tuning the metal-insulator transition temperature by the controlled generation of oxygen vacancies on La0.7Ca0.3MnO3−x epitaxial thin films. *Journal of Alloys and Compounds* 989 (2024): 174321.
100. Verhage, M., and Flipse, K. Order in disorder: Oxygen vacancy driven electronic phase separation of LaNiO3−x epitaxial thin film surface investigated by scanning probe microscopy. *arXiv preprint* arXiv:2410.06598 (2024).
101. Blume, R., Calvet, W., Ghafari, A., Mayer, T., Knop-Gericke, A., and Schlögl, R. Structural and chemical properties of NiOx thin films: Oxygen vacancy formation in O2 atmosphere. *ChemPhysChem* 24.23 (2023): e202300231.

5 Defects in Thin Films – Part II

Adeeba Mirza and Abhilasha Mishra

5.1 INTRODUCTION

Thin film coatings are an important aspect in various industries, such as electronics, optics, sensors, photovoltaics, and protective coatings for various mechanical and other applications. The properties and functions of these thin films are controlled and tailored by film thickness. Defects in thin films have a significant effect on their structural and mechanical performance and reliability [1–4]. Often, these defects are attributed to thermal, mechanical, or environmental factors. Defects in thin films can be categorized into point, line, planar, and volumetric defects. These defects have varied adverse effects on the structural, electrical, optical, and mechanical properties of thin films, leading to the failure of MEMS components, solar cell, biomedical components, optical coatings, etc. Therefore, a proper characterization of thin films is imperative to not only understand the origin and their effect on thin films but also to optimize the deposition parameters during deposition techniques such as PVD, CVD, and sputtering among others. A wide range of techniques are used to understand the defects in thin films [5–7]. The surface of thin films is characterized using atomic force microscopy and scanning electron microscopy to evaluate the topographical information such as pin holes, blowholes, cracks, and roughness. The texture of thin films is an important attribute that dictates their functionalities. The details of their microstructure are analysed using transmission electron microscopy, which gives information on grain dislocation and grain boundaries. Also, X-ray diffraction technique is used to analyse the crystallographic orientation, phase identification, and residual stress [5–6].

The chemical analysis of thin films is done using techniques such as X-ray photoelectron spectroscopy (XPS) and auger electron spectroscopy (AES). Further, Raman spectroscopy, ellipsometry, and secondary ion mass spectroscopy (SIMS) are valuable methods to determine the optical constants, vibrational modes, and depth profiling of the elemental distribution in thin films [2, 8, 9]. A detailed and comprehensive characterization of thin film coatings is important for ensuring the quality of thin films, guiding optimization of the process parameters during deposition, thereby facilitating development of high-performance material. With the advent of miniaturization and multifunctionality in engineering components, understanding the properties of thin films in a holistic manner using proper characterization techniques is central to material science and engineering research.

DOI: 10.1201/9781003750659-5

5.2 CHARACTERIZATION TECHNIQUE FOR DEFECT ANALYSIS

To ensure quality and performance, characterizing defects in thin films is crucial. Various advanced techniques are employed for this purpose [10]. AFM stands out as a high-resolution imaging device that helps in the measurement of surface roughness and the determination of various defects on the surface. SEM offers detailed surface morphology imaging, identifying surface defects, grain boundaries, and other microstructural features, while TEM provides high-resolution imaging and analysis of internal structures, detecting crystallographic defects like dislocations and stacking faults [11]. X-ray diffraction (XRD) analyses crystallographic structure and orientation, identifying lattice strain, dislocations, and phase composition. Raman spectroscopy probes vibrational modes, providing insights into chemical composition, stress, and phase transitions [12].

For surface chemical composition, X-ray photoelectron spectroscopy (XPS) is greatly successful, identifying surface contaminants and chemical bonding environments. Time-of-flight secondary ion mass spectrometry (SIMS) offers surface composition and molecular structure analysis, detecting trace impurities and surface contamination [13]. Ellipsometry measures film thickness and optical properties, detecting variations and interface roughness. Photoluminescence spectroscopy examines different properties and helps identify defects related to electronic states and recombination centres, while Fourier transform infrared spectroscopy probes vibrational modes to identify chemical bonds and impurities [14].

Electrochemical analysis method, such as cyclic voltammetry, investigates electrochemical properties, identifying defects that impact conductivity and corrosion resistance [15]. Positron annihilation spectroscopy (PAS) is used for analysing vacancy-type defects, detecting vacancies, voids, and defect clusters. Also, ion beam analysis (IBA) techniques, like Rutherford backscattering spectrometry (RBS), measure composition and thickness and detect compositional variations and interface quality [16]. These techniques, often used in combination, provide a brief understanding of defects in thin films, facilitating their optimization for various applications.

5.3 SCANNING ELECTRON MICROSCOPY (SEM)

SEM is a powerful and flexible method that is widely applied for identification and analysis of defects in thin film coatings (TFCs). SEM provides high-resolution images and delivers information about the surface morphology of thin films [17]. SEM works by scanning a beam of high-energy electrons across the sample surface. As the electron beam hits the sample, it generates different signals such as secondary electrons and characteristic rays. The signals are detected and processed to generate high-resolution images and give compositional data. Scanning electron microscopes are highly valued tools with high resolution, enabling scientists to study the smallest feature and faults on the surface of a sample to as little as a few nanometres. They are greatly capable of capturing the depth of field and rendering into sharp 3D imagery. This functionality makes SEM ideal for examining the topography and roughness of a varied range of materials, from TFC to more robust specimens [18]. These features make these tools a valuable asset in numerous fields.

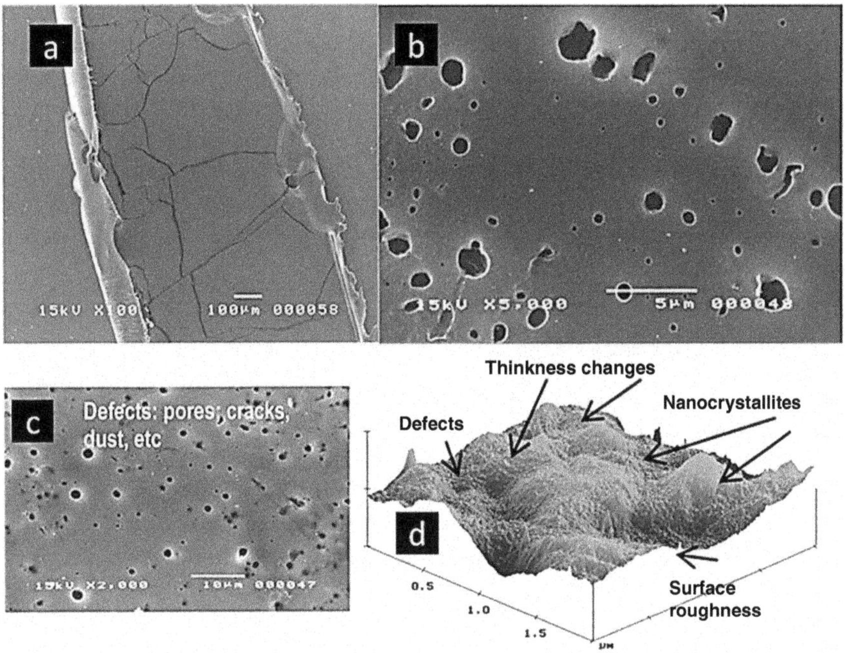

FIGURE 5.1 SEM images (a) and (b) showing the micro cracks and pores on coated glass respectively. Image (c) shows a highly porous surface and image (d) gives information about AFM of thin film exhibiting average roughness and singular morphology (Reproduced with permission [19]).

5.3.1 Types of Defects Analysed by SEM

SEM is highly effective in identifying surface defects such as scratches, pits, pinholes, and voids [20] (See Figure 5.1). These defects are critical in determining the mechanical and protective properties of TFC. These faults tend to greatly influence the performance and functionality of the film. Moreover, it can reveal structural faults like grain boundaries, which tend to readily deteriorate the quality of the films. SEM further aids in film uniformity evaluation, enabling one to detect thickness variances and delamination. Lastly, SEM can detect foreign particles and impurities on the surface of the film, all of which can cause failure or performance issues [21]. SEM provides several imaging modes to analyse defects in thin films. Secondary electron imaging (SEI) gives high-resolution surface information, which is best suited to detect small defects such as scratches or pits. Backscattered electron imaging (BEI) shows compositional contrasts by indicating heavier elements, useful for detecting inclusions or non-uniform material distribution.

Energy-dispersive X-ray spectroscopy (EDXS), usually utilized in combination with SEM, facilitates identification of the precise elements with defects contaminants and inclusions for thorough analysis [22]. Preparation of the sample is essential for obtaining good-quality SEM images. Careful sample preparation images are necessary for SEM analysis to produce the best results. Conductive films, such as gold or carbon, could be necessary for non-conductive thin films to avoid electrical charging during exposure to the electron beam [23]. For the study of internal defects or at the interface between various layers within the film, specialized methods such as FIB (focused ion beam) milling could be applied to prepare thin film cross sections. The advantages of SEM analysis are high resolution and depth of field and detailed topographical and compositional information. However, it requires vacuum conditions, which may not be suitable for all samples. Also, non-conductive samples need special preparation and limited penetration depth, making it primarily a surface analysis technique.

5.4 X-RAY DIFFRACTION (XRD)

This method is used to study the defects in thin films type of coatings. It analyses the diffraction patterns obtained when X-rays interact with the crystal lattice [24]. X-rays are made to fall on the sample TFC at different angles, and the resulting diffraction pattern is measured. This pattern is made up of peaks that relate to the angles where X-rays are diffracted by the crystal lattice planes. Based on these diffraction peaks, interesting information regarding the crystal structure and thin-film defects can be gathered [25].

XRD is a powerful method to study thin-film defects through the utilization of their crystallographic fingerprint [26]. It can yield a lot of information, such as the crystal structure of the film, lattice parameters, and phase content. Any departure from a perfect crystalline structure, detected by peak shifts or broadening, can be an indication of the existence of defects. It may also uncover information regarding the microstructure of the film, including grain sizes, where irregularities may represent defects such as grain boundaries or dislocations. In addition, XRD may evaluate residual stresses in the TFC, another condition that can lead to defects [27].

5.4.1 Types of Defects Analysed by XRD

Dislocations, such as misplaced atoms, result in distortions that are seen in shifted or broadened diffraction peaks. The analysis of these shifts can even identify the type and density of dislocations (edge or screw) present [28]. Stacking faults, where the atomic order is disturbed, are another offender. XRD identifies them by the presence of new peaks and changes in the intensity of existing peaks relative to a perfect crystal structure. Lastly, grain boundaries, where crystal orientation alters, can be detected through variations in peak shape and intensity. This information enables the quantification of grain size variation and lattice mismatch between adjacent grains and the location of these boundaries. XRD analysis demands a thoughtful choice of the substrate upon which thin film deposition is done [29]. In the ideal case, the substrate should absorb and scatter as little X-rays as possible to produce an unobscured diffraction pattern due to the film. Moreover, to investigate the crystallographic texture or preferred orientation of the film, the substrate may have to be selected with a particular crystallographic orientation that favours the preferable growth mode in the thin film. The advantages of XRD analysis are (i) detailed crystallographic information about thin films, (ii) non-destructive and capable of analysing crystalline phases and defects, and (iii) it can be used for mixed methods analysis. However, it has a few limitations such as (i) limited sensitivity to defects in amorphous or highly disordered materials, (ii) requires careful sample preparation and substrate considerations, and (iii) some defects may not be directly detectable and may require complementary techniques for confirmation.

5.5 TRANSMISSION ELECTRON MICROSCOPY (TEM)

TEM is a versatile tool which has extensive applications in the identification and characterization of defects in TFC with atomic resolution [30]. TEM is based on the transmission principle of a well-focused electron beam through a thin sample, which depicts the atomic-scale morphology and defects [31]. In this technique, electrons are accelerated and concentrated on the thin film sample, which is mounted as a thin cross section or thin foil. The electrons interact with the sample atoms, experiencing scattering and diffraction. The transmitted electrons create an image that is magnified and projected on a fluorescent screen or captured by a digital camera [32].

TEM meticulously reveals defects at the atomic scale. With its unparalleled resolution, TEM enables scientists to see directly individual atomic columns in the crystal lattice of thin films, displaying defects in detail [33]. In addition to defect imaging, it provides a wealth of crystallographic information. It can determine the exact dimensions of the film's crystal lattice. Different imaging modes, employing varying resolutions and lighting conditions, provide versatile views of defects like dislocations, stacking faults, vacancies, and interfaces within the film. Additionally, electron diffraction patterns can offer further insights into the crystallographic properties of the thin film. With these capabilities, TEM becomes a powerful tool for comprehensively characterizing defects and their impact on thin film properties.

5.5.1 Types of Defects Analysed by TEM

Taking TEM a step further, we can pinpoint specific types of defects with incredible detail. Dislocations, appearing as line-like disruptions in the lattice, are directly imaged by TEM [34]. By analysing these lines, scientists can differentiate between edge, screw, or mixed dislocations based on their atomic arrangement (Burger's vector) and direction [35]. Stacking faults and vacancies (See Figure 5.2), regions where the atomic order is disrupted, are also visualized by TEM, allowing for identification of their type (intrinsic or extrinsic) and their density within the film. Grain boundaries, the transition zones between differently oriented grains, are revealed by abrupt changes in crystal orientation. TEM can even characterize these boundaries by measuring the misorientation angle, type (e.g., tilt or twist), and any associated defect structures present [36]. Also, it allows for the detection of point defects like vacancies, interstitials, and impurity atoms, all appearing as distinct perturbations within the crystal lattice.

While TEM offers unparalleled defect analysis, careful sample preparation is crucial for optimal results [38]. To achieve electron transparency, thin film samples are sculpted into ultrathin foils using techniques like focused ion beam milling or mechanical thinning. Additionally, specific crystallographic planes or grain

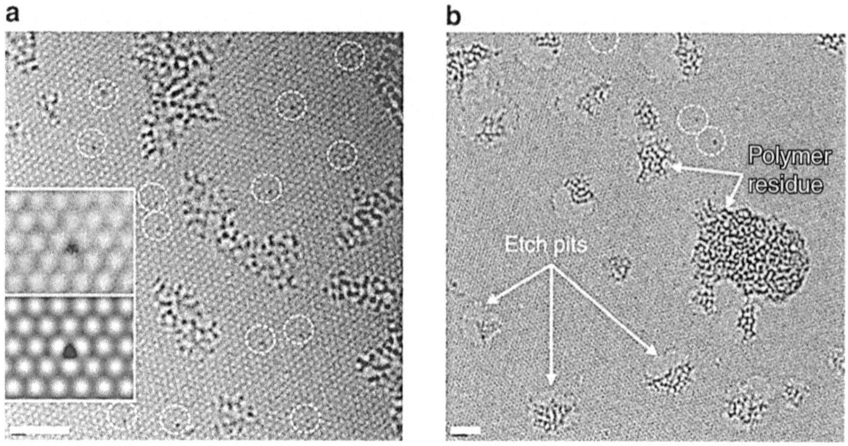

FIGURE 5.2 Vacancy-type defect analyses by TEM (Reproduced with permission [37]).

boundaries of interest might require targeted orientation during sample preparation for focused defect analysis. Finally, minimizing artefacts caused by surface amorphization (loss of crystallinity), ion beam damage from the thinning process, and specimen contamination is essential for obtaining high-quality TEM images that accurately represent the defects within the thin film [39]. The advantages of TEM analysis are (i) atomic-level resolution and detailed imaging of defects, (ii) ability to visualize crystallographic structure, defects, and interfaces and (iii) provides complementary information to other techniques like XRD and SEM. However, the limitations are (i) sample preparation can be time-consuming and challenging, particularly for thin films, (ii) electron beam damage may occur, limiting observation time and resolution, and (iii) it requires expensive instrumentation and expertise for operation and interpretation.

5.6 ATOMIC FORCE MICROSCOPY (AFM)

AFM is a strong and flexible method for the detection and analysis of thin-film defects with high-resolution imaging ability [40]. AFM operates based on the connection between a sharp probe tip and the thin-film surface, giving details at the nanometre level [41]. In AFM, a probe tip that is very sharp and supported on a cantilever is brought near to the surface of the thin film. When the probe tip is scanning across the surface, the probe tip experiences forces like van der Waals forces and electrostatic forces which make the plank deflect [42]. The deflection is sensed by a laser beam reflected off the back of the cantilever, permitting accurate mapping of the surface topography. AFM has an incredibly strong toolbox to call on when investigating thin film defects. It has fantastic resolution, obtaining sub-nanometre resolution laterally and atomic-scale vertical precision. This means it has incredibly detailed pictures of defects, surface roughness, and topographical features where they exist within the film. Its sensitivity makes it capable of picking up minute nanoscale defects, scratches, and surface contaminants. Moreover, AFM offers a key 3D view. It can image surface morphology and defects not only laterally but also in height, providing a more complete picture [43]. The abilities of AFM go beyond imaging. It can make force spectroscopy measurements, in effect sensing the surface. This enables researchers to govern the mechanical properties of the TFC and even sense changes in stiffness and friction that may be related to the presence of defects.

5.6.1 Types of Defects Analysed by AFM

AFM can identify multiple surface defects present in thin films, such as scratches, pits, and surface contaminations [44]. The defects can be seen in high-resolution images and quantified with respect to their size and distribution. This technique can even find nanoscale characteristics such as grain boundaries and surface roughness in the thin films. Such characteristics give good information regarding the microstructural characteristics and defect distribution in the film. AFM-based force spectroscopy methods can investigate mechanical faults like adhesion failures, delamination, and variations in surface roughness related to faults in thin films [45]. Sample preparation is essential for the acquisition of good AFM images of thin film defects [46]. Thin film samples need to be deposited onto smooth substrates with low surface roughness to enable accurate topographical imaging. Samples need to be cleaned before imaging to eliminate surface debris and contaminants that can disrupt AFM measurements [47]. Certain samples need surface conditioning treatments like chemical treatment or annealing to improve surface quality and reduce surface defects. The advantages of AFM analysis are (i) high-resolution imaging of surface morphology and defects, (ii) surface sensitivity and 3D imaging capabilities, and (iii) versatility and compatibility with various sample types and environments. However, its limitations are (i) its sole function is surface imaging and is unable to probe internal defects, (ii) imaging speed may be slower compared to other techniques, and (iii) it requires careful sample preparation and probe selection for accurate measurements.

5.7 FACTORS AFFECTING DEFECTS FORMATION

5.7.1 Material Properties

Material properties have an overwhelming impact on defect formation in thin films due to the unique structural and mechanical characteristics of thin film materials [48]. Here's how various material properties influence (See Figure 5.3) defect formation in thin films.

i) Crystal Structure and Grain Size

The crystal structure of the material imposes its defect formation behaviour. In thin films, where the volume is inadequate, defects such as dislocations, vacancies, and grain boundaries can extensively affect mechanical and electrical properties [49]. Polycrystalline films may contain grain boundaries where defects tend to accumulate, impacting film integrity and performance. Thin films comprise grains varying in dimensions from nanometres to micrometres. Smaller grain sizes generally lead

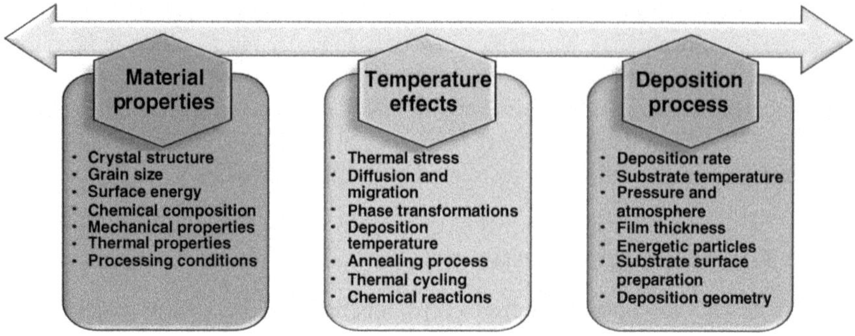

FIGURE 5.3 Factors affecting defects formation.

to a higher density of grain boundaries, which serve as sinks for defects [50]. Grain boundaries can act as barriers to dislocation motion, leading to strain accumulation and defect nucleation. Controlling grain size is crucial for minimizing defect formation and enhancing mechanical strength in thin films.

ii) Surface Energy and Chemical Composition

Material surface energy influences defect formation on the thin film surface. High surface energy materials are more prone to defects such as islands, voids, and cracks due to increased surface mobility [51]. Surface defects can raise into the bulk of the film, affecting its mechanical and optical properties. The chemical composition of thin films affects defect formation through the introduction of point defects, such as vacancies and interstitials. Variations in stoichiometry or impurity content can create structural defects like dislocations, stacking faults, and phase boundaries. Chemical reactions linking the thin film and the substrate or ambient environment can also lead to defect formation and degradation of film properties.

iii) Mechanical and Thermal Properties

Mechanical properties such as hardness, elasticity, and ductility determine defect formation by affecting how films react to external stress [52]. Low-ductility films are prone to delamination and cracking, resulting in defect formation along interfaces and grain boundaries. Mechanical stress applied during film deposition or post-processing can introduce defects such as surface roughness and dislocations. Thermal properties like thermal expansion coefficient and thermal conductivity perturb effect formation in temperature-strayed thin films [53]. Thermal expansion coefficient mismatch between the substrate and film may create thermal stresses and thus lead to defect formation such as cracks and voids. Thermal cycling also facilitates diffusion-related defect formation mechanisms.

iv) Processing Conditions and Temperature Effect

Deposition method, pressure, and environment are essential factors in the formation of defects in thin film creation [54]. Inadequate processing conditions may result in defects such as adhesion failure, porosity, and non-uniform thickness of the film, undermining the quality and performance of the film. Both these material characteristics and defect formation have a significant connection towards optimizing thin film deposition processes and the stability of thin film devices in different applications, ranging from electronics to optics and coatings. Temperature plays a vital role in defect formation in TFC through various mechanisms. The effect of temperature in defect formation in thin films is illustrated in Figure 5.3 [55].

v) Thermal Stresses Diffusion and Migration

Thermal expansion coefficient differences between the TFC (thin film) and the substrate can cause thermal stresses when the temperature varies. These stresses can initiate the development of cracks and other structural imperfections. For instance, when cooling down from the deposition temperature, tensile stresses could be developed if the film shrinks more than the substrate, leading to cracking.

Higher temperatures can enhance the mobility of atoms and vacancies in the thin film, enhancing diffusion and migration [56]. Increased mobility can cause defect healing, where the number of existing defects such as vacancies and dislocations is minimized. It can also cause the generation of new defects such as voids and hillocks if diffusion is non-uniform or there are impurities.

vi) Phase Transformations and Deposition Temperature

Temperature variations can cause phase transitions in thin films, which can create defects [57]. For example, a temperature variation can cause a transition phase change from amorphous to crystalline. These transitions can create defects such as dislocations, stacking faults, and grain boundaries because of variations in atomic structures and lattice parameters between phases. The deposition temperature set during thin film deposition significantly affects its microstructure and defect density. Increased deposition and temperature tend to enhance crystallinity and lower defect density by allowing adequate energy for atoms to reorganize into configurations of lower energy [58]. Lower deposition temperatures can lead to an increased defect density of types such as vacancies, interstitials, and grain boundaries owing to restricted atomic mobility.

vii) Annealing Processes and Thermal Cycling

Following the deposition thermal treatments, annealing can influence defect formation. Annealing at suitable temperatures can decrease the density of defects by allowing atoms to migrate and remove defects such as dislocations and grain boundaries through mechanisms like recrystallization and grain growth [59]. But, if the annealing temperature is too high, it may lead to adverse effects such as excessive film cracking, grain growth, and introducing new defects. Recurring heating and cooling cycles can produce cyclic thermal stresses, leading to fatigue and the accumulation of defects over time. It can cause defects such as microcracks, voids, and delamination, mainly in the thin films with different thermal expansion coefficients compared to their substrates [60].

viii) Chemical Reactions

High temperatures promote reactions between thin films and their surroundings, causing defects to form. For example, oxidation, interdiffusion, or chemical reactions at the film–substrate interface can introduce defects like voids, intermetallic phases, or composition gradients, degrading the film's integrity and performance. Temperature significantly influences defect formation in thin films through mechanisms like thermal stress, diffusion, phase transformations, deposition conditions, annealing, thermal cycling, and chemical reactions [61]. Controlling the temperature during and after the deposition processing is crucial for minimizing defects and improving the properties of TFC.

ix) Deposition Process and Deposition Technique

The deposition process has a profound impact on the formation of defects in thin films. Different deposition methods and conditions can introduce various types of defects, influencing the film's structural, mechanical, and functional properties [62]. The effect of deposition process on the origination of defects in thin films is shown in Figure 5.3. Thin films are susceptible to defects depending on the chosen deposition technique. Techniques like physical vapour deposition (PVD) and sputtering can introduce voids, pinholes, and stress due to their high-energy processes. Chemical vapour deposition (CVD) offers more control but can still result in defects like voids and dislocations. Atomic layer deposition (ALD) is known for minimal defects, but incomplete reactions can introduce some. Molecular beam epitaxy (MBE) offers exceptional control, but careful management is needed to avoid surface roughness and dislocations.

x) Deposition Rate and Substrate Temperature

Beyond the technique, various factors within the deposition process itself influence defect formation. Deposition rate plays a key role. A very high rate can lead to rough surfaces, uneven thickness, and more defects due to limited time for atoms to arrange themselves properly. Conversely, a slower rate allows for better crystal formation and fewer defects.

Substrate temperature is another critical factor. Higher temperatures enhance atomic mobility, promoting good crystal formation and reducing defects. However, excessively high temperatures can cause unwanted material diffusion at the interface, introducing new defects. On the other hand, low temperatures hinder atomic movement, potentially resulting in more defects within the film.

xi) Pressure and Atmosphere

Several other factors also influence defect formation. Pressure and gas composition in the chamber affect film growth. Low-pressure environments lead to fewer defects, while higher pressure can cause more defects. Additionally, film thickness itself can play a role, with very thin films potentially having more defects due to limited atomic mobility [63]. By carefully controlling all these parameters, scientists can achieve high-quality thin films with minimal defects. The deposition process, including the technique, rate, temperature, pressure, atmosphere, thickness, particle energy, substrate preparation, and deposition geometry, critically influences the formation and density of defects in thin films. Adjusting these factors can result in high-quality thin films with minimal defects, produced for specific applications.

5.8 MITIGATION STRATEGIES FOR DEFECT REDUCTION

Reducing defects in TFC is crucial for improving their functionality, reliability, and performance across various applications [64]. Mitigation strategies aim to minimize defect formation during thin film deposition, processing, and post-treatment stages [65]. Here are some effective strategies for defect reduction.

5.8.1 Optimized Deposition Parameters and Improved Substrate Cleaning

Fine-tuning deposition parameters can minimize defect formation. Maintaining uniform deposition conditions and optimizing film growth kinetics help promote epitaxial growth and reduce defects like dislocations and grain boundaries [66, 67]. Thorough cleaning and surface preparation of substrates prior to thin film deposition help remove contaminants, impurities, and surface defects that can act as nucleation sites for defect formation. Techniques such as solvent cleaning, ultrasonic cleaning, and plasma cleaning enhance substrate cleanliness and adhesion, reducing defect density [68].

5.8.2 Controlled Gas Atmosphere and Precursor Purity and Quality

Maintaining a controlled gas atmosphere during deposition processes, such as CVD or PVD, helps prevent oxidation, contamination, and chemical reactions that contribute to defect formation [69]. Inert gas purging or the use of protective atmospheres can minimize oxygen and moisture exposure, reducing surface defects and film degradation [70]. Using high-purity precursor materials and deposition sources minimizes the introduction of impurities and defects during TFC synthesis [71]. Purification techniques such as sublimation, distillation, and chemical purification ensure the quality and consistency of precursor materials, resulting in defect-free thin films [72].

5.8.3 Substrate Temperature Control and Surface Passivation

Precise control of substrate temperature during deposition helps optimize thin film nucleation, growth kinetics, and crystallinity, reducing defects like vacancies, dislocations, and grain boundaries [73]. Temperature ramping, thermal annealing, and substrate preheating minimize thermal stress and promote defect annealing during thin film growth.

Surface passivation techniques, such as chemical treatments, surface coatings, and self-assembled monolayers (SAMs), protect thin film surfaces from contamination, oxidation, and chemical reactions that lead to defect formation [74]. Passivation layers enhance surface stability, adhesion, and corrosion resistance, reducing defect density and improving film quality.

5.8.4 Post-Deposition Annealing and In Situ Monitoring and Control

Thermal annealing treatments performed after thin film deposition help eliminate defects, enhance crystallinity, and improve film properties [75]. Annealing at optimized temperatures and durations facilitates defect diffusion, grain growth, and stress relaxation, leading to defect reduction and improved film performance.

Real-time monitoring techniques, such as spectroscopic ellipsometry, quartz crystal microbalance (QCM), and in situ stress measurements, enable the continuous monitoring and control of thin film deposition processes [76]. To minimize defects and optimize film quality during growth, feedback control systems are used in real time.

5.8.5 Advanced Deposition Techniques and Defect Analysis and Characterization

Utilizing advanced deposition techniques, including ALD, MBE, and PLD, offers exact control on the thickness, composition, and crystallinity, minimizing defects or achieving high-quality TFC with atomic-scale precision [77]. Regular defect analysis and characterization using techniques such as SEM, TEM, XRD, and AFM help identify defect types, locations, and densities [78]. A detailed discussion on defect analysis has been shown through flow chart in Figure 5.4. Understanding defect formation mechanisms guides the optimization of mitigation strategies for defect reduction in thin films [79]. By implementing these mitigation strategies and optimizing thin film deposition processes, manufacturers can effectively reduce defect density, improve film quality, and enhance the performance of thin film-based devices and coatings for various industrial and technological applications.

5.8.6 Future Directions and Opportunities and Emerging Materials and Nanostructures

Future directions and opportunities in the field of thin films present exciting prospects for advancements in materials science, device technology, and industrial applications [80]. Continued exploration of novel materials, such as two-dimensional materials (e.g., graphene, transition metal dichalcogenides), perovskites, and quantum dots, offers opportunities for developing advanced thin films [81]. Integration of nanostructures, nanocomposites, and heterostructures allows the design of tailored TFC materials with enhanced performance for electronics, photonics, and energy applications [82].

5.8.7 Thin Film's Functional Devices

Advancements in TFC fabrication techniques and materials design cover the path for the progress of functional devices with improved performance and functionality [83]. Thin film-based devices, including flexible electronics, wearable sensors, photovoltaics, and transparent conductive coatings, offer opportunities for diverse applications in electronics, healthcare, energy harvesting, and in smart technologies [84].

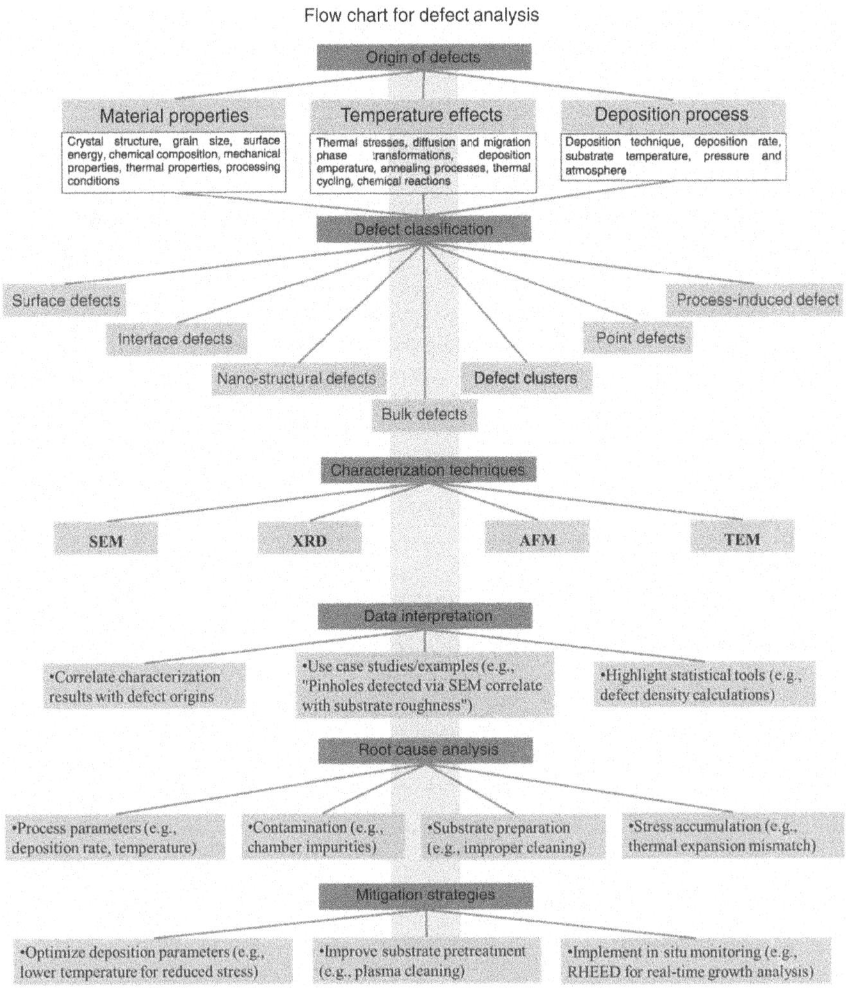

FIGURE 5.4 Flow chart for defect analysis in thin films.

5.8.8 Thin Film Deposition Technologies and Multifunctional Thin Film Coatings

Innovation in TFC deposition technologies, include ALD, CVD, and inkjet printing, allows exact control on the TFC thickness and morphology [85]. Integration of additive manufacturing techniques and roll-to-roll processing facilitates scalable production of thin films with high throughput, cost-effectiveness, and customizability for industrial-scale applications [86]. Development of multifunctional TFCs with integrated properties, including self-cleaning, anti-fogging, and antimicrobial functionalities, addresses diverse industry needs in automotive, aerospace, healthcare, and infrastructure sectors [87]. Functional coatings offer opportunities for improving product performance, durability, and sustainability while reducing maintenance costs and environmental impact [88].

5.8.9 Smart and Adaptive Materials and Advanced Characterization Techniques

Research in smart and adaptive thin film materials, including shape memory alloys, piezoelectric films, and phase-change materials, enables the design of responsive surfaces and stimuli-responsive devices [89]. Integration of sensing, actuation, and control functionalities into thin film coatings opens up opportunities for applications in robotics, smart textiles, biomedical devices, and environmental monitoring [90]. Advancements in characterization techniques, such as in situ and operando methods, electron microscopy, spectroscopy, and machine learning-based analytics, provide deeper insights into thin film structure–property relationships, defect formation mechanisms, and performance optimization strategies [91]. Integration of multi-modal and multi-scale characterization approaches enhances our understanding of thin film materials and accelerates materials discovery and development [92].

5.8.10 Sustainable Thin Film Technologies

Growing emphasis on sustainability and environmental stewardship drives research and innovation in sustainable thin film technologies, including eco-friendly materials, green synthesis methods, and energy-efficient deposition processes [93]. Adoption of circular economy principles and life cycle assessments promotes the growth of environmentally friendly TFC solutions for a huge range of industries applications [94].

5.8.11 Collaborative Research, Industry Partnerships, and Global Market Expansion

Collaboration between academia, research institutions, and industry stakeholders fosters interdisciplinary research, technology transfer, and commercialization of thin film innovations [95]. Joint initiatives, consortia, and public-private partnerships accelerate the translation of fundamental research into practical applications, driving economic growth, competitiveness, and societal impact.

Growing request for TFC-based products and technologies across diverse sectors, including electronics, healthcare, energy, and infrastructure, fuels market expansion and investment opportunities [96]. Emerging markets in Asia-Pacific, Latin America, and Africa offer untapped potential for thin film applications, driving global market growth and diversification. The future directions and opportunities in thin films encompass a wide spectrum of research, development, and commercialization activities, spanning materials innovation, device technology, process optimization, and sustainability initiatives [97]. By leveraging emerging trends, collaborative partnerships, and market-driven solutions, the thin film industry is poised for continued growth, also societal impact and innovation in the years to come [98].

5.9 CONCLUSION AND FUTURE PROSPECTS

In conclusion, understanding the defects in thin films is imperative to evaluate their performance and reliability for various engineering applications. The comprehensive review in this chapter has explored different defects that occur in thin films, including point, line, planar, and volumetric defects. These defects can be originated during deposition or post processing of thin film coatings. These defects have a significant impact on their structural make-up, chemical, physical, optical, electrical, and mechanical properties among others. To reduce the defects in TFCs, it is imperative that suitable characterization techniques are employed to understand the information about the origin of these defects.

Additionally, continuous advancements in materials synthesis, deposition methods, and characterization techniques will provide new avenues for defect analysis and control in TFC. The development of next-generation thin film materials and devices is made possible by emerging technologies like ALD, additive manufacturing, and in situ monitoring, which provide precise control over TFC properties and defect reduction techniques. Collaboration between government, business, and academic institutions will be essential as this field of study develops to spur innovation, hasten the adoption of new technologies, and solve practical problems. We can unlock the full potential of TFC materials for a variety of applications by improving our knowledge of thin-film defects and creating efficient mitigation techniques.

REFERENCES

1. Vyas, S. A short review on properties and applications of zinc oxide based thin films and devices: ZnO as a promising material for applications in electronics, optoelectronics, biomedical and sensors. *Johnson Matthey Technology Review* 64.2 (2020): 202–218.
2. Li, W., Shi, J., Zhang, K. H., and MacManus-Driscoll, J. L. defects in complex oxide thin films for electronics and energy applications: Challenges and opportunities. *Materials Horizons* 7.11 (2020): 2832–2859.
3. Glynn, C., and O'Dwyer, C. Solution processable metal oxide thin film deposition and material growth for electronic and photonic devices. *Advanced Materials Interfaces* 4.2 (2017): 1600610.
4. Bart, J. J. Analysis of deposited glass layer defects. In *13th international reliability physics symposium*. IEEE, 1975: 128–135.
5. Pandey, A., Dalal, S., Dutta, S., and Dixit, A. Structural characterization of polycrystalline thin films by X-ray diffraction techniques. *Journal of Materials Science: Materials in Electronics* 32 (2021): 1341–1368.
6. Vahidi, H., Syed, K., Guo, H., Wang, X., Wardini, J. L., Martinez, J., and Bowman, W. J. A review of grain boundary and heterointerface characterization in polycrystalline oxides by (scanning) transmission electron microscopy. *Crystals* 11.8 (2021): 878.
7. Yan, Y., Wang, Y., Zhou, P., Huang, N., and Guo, D. Near-field microscopy inspection of nano scratch defects on the monocrystalline silicon surface. *Precision Engineering* 56 (2019): 506–512.
8. Richter, A., Czerwosz, E., Dłużewski, P., Kozłowski, M., and Nowicki, M. Nanoindentation of heterogeneous carbonaceous films containing Ni nano-crystals. *Micron* 40.1 (2009): 94–98.
9. Evans, P. J., Murai, Y., Lindsay, M. J., Davis, J., and Triani, G. Properties of ALD Al_2O_3 protective coatings. *American Vacuum Society* (2010).

10. Abou-Ras, D., Kirchartz, T., and Rau, U. (Eds.). *Advanced characterization techniques for thin film solar cells*. Vol. 2. Wiley-VCH, 2011.
11. Inkson, B. J. Scanning electron microscopy (SEM) and transmission electron microscopy (TEM) for materials characterization. In *Materials characterization using nondestructive evaluation (NDE) methods*. Woodhead Publishing, 2016: 17–43.
12. Venkataraman, K., Kropf, A. J., Segre, C. U., Jia, Q. X., Goyal, A., Kang, B. W., and Maroni, V. A. detection of interfacial strain and phase separation in MBa2Cu3O7–x thin films using Raman spectroscopy and X-ray diffraction space mapping. *Physica C: Superconductivity* 402.1–2 (2004): 1–16.
13. Lockyer, N. P., Aoyagi, S., Fletcher, J. S., Gilmore, I. S., van der Heide, P. A., Moore, K. L., and Weng, L. T. Secondary ion mass spectrometry. *Nature Reviews Methods Primers* 4.1 (2024): 32.
14. Zhang, J. Z. *Optical properties and spectroscopy of nanomaterials*. World Scientific, 2009.
15. Cesiulis, H., Tsyntsaru, N., Ramanavicius, A., and Ragoisha, G. The study of thin films by electrochemical impedance spectroscopy. In *Nanostructures and thin films for multifunctional applications: Technology, properties and devices*. Cham: Springer International Publishing, 2016: 3–42.
16. Zhang, Y., Debelle, A., Boulle, A., Kluth, P., and Tuomisto, F. Advanced techniques for characterization of ion beam modified materials. *Current Opinion in Solid State and Materials Science* 19.1 (2015): 19–28.
17. Baer, D. R., and Thevuthasan, S. Characterization of thin films and coatings. In *Handbook of deposition technologies for films and coatings*. William Andrew Publishing, 2010: 749–864.
18. Kassim, A., Ho, S. M., Tan, W. T., and Atan, S. Chemical bath deposition of ZnSe thin films: SEM and XRD characterization. *European Journal of Applied Sciences* 3.3 (2011): 113–116.
19. Vasconcelos, H. C., and Pinto, A. S. Fluorescence properties of rare-earth-doped sol-gel glasses. In *Recent applications in sol-gel synthesis*. IntechOpen, 2017: 81–107.
20. Panjan, P., Čekada, M., Panjan, M., and Kek-Merl, D. Growth defects in PVD hard coatings. *Vacuum* 84.1 (2009): 209–214.
21. Rancourt, J. D., Brooks, J., Mecham, S., Sentman, A., Starr, B., and Todd, J. Failure, defect, and contaminant analysis. *Comprehensive Analytical Chemistry* 53 (2008): 607–674.
22. Alqaheem, Y., and Alomair, A. A. Microscopy and spectroscopy techniques for characterization of polymeric membranes. *Membranes* 10.2 (2020): 33.
23. Echlin, P. *Handbook of sample preparation for scanning electron microscopy and X-ray microanalysis*. Springer Science & Business Media, 2011.
24. Raghothamachar, B., Dudley, M., and Dhanaraj, G. X-ray topography techniques for defect characterization of crystals. In *Springer handbook of crystal growth*. Springer, 2010: 1425–1451.
25. Widjonarko, N. E. Introduction to advanced X-ray diffraction techniques for polymeric thin films. *Coatings* 6.4 (2016): 54.
26. Khan, H., Yerramilli, A. S., D'Oliveira, A., Alford, T. L., Boffito, D. C., and Patience, G. S. Experimental methods in chemical engineering: X-ray diffraction spectroscopy – XRD. *The Canadian Journal of Chemical Engineering* 98.6 (2020): 1255–1266.
27. Fitzpatrick, M. E., Fry, A. T., Holdway, P., Kandil, F. A., Shackleton, J., and Suominen, L. determination of residual stresses by X-ray diffraction. NPL Publication, 2005.
28. Meng, F. E. I., Morin, S. A., Forticaux, A., and Jin, S. Screw dislocation driven growth of nanomaterials. *Accounts of Chemical Research* 46.7 (2013): 1616–1626.
29. Birkholz, M. *Thin film analysis by X-ray scattering*. John Wiley & Sons, 2006.

30. Lin, Y., Zhou, M., Tai, X., Li, H., Han, X., and Yu, J. Analytical transmission electron microscopy for emerging advanced materials. *Matter* 4.7 (2021): 2309–2339.
31. Javed, Y., Ali, K., Akhtar, K., Jawaria, Hussain, M. I., Ahmad, G., and Arif, T. TEM for atomic-scale study: Fundamental, instrumentation, and applications in nanotechnology. In *Handbook of materials characterization*. Cham: Springer International Publishing, 2018: 147–216.
32. Carter, C. B., and Williams, D. B. (Eds.). *Transmission electron microscopy: Diffraction, imaging, and spectrometry*. Springer, 2016.
33. Dolabella, S., Borzì, A., Dommann, A., and Neels, A. Lattice strain and defects analysis in nanostructured semiconductor materials and devices by high-resolution X-ray diffraction: Theoretical and practical aspects. *Small Methods* 6.2 (2022): 2100932.
34. Walter, T., and Gerthsen, D. TEM analysis of epitaxial semiconductor layers with high stacking fault densities considering artifacts induced by the cross-section geometry. *Ultramicroscopy* 81.3–4 (2000): 279–288.
35. Matsuhata, H., Yamaguchi, H., and Ohno, T. Analysis of contrasts and identifications of Burgers vectors for basal-plane dislocations and threading edge dislocations in 4H-SiC crystals observed by monochromatic synchrotron X-ray topography in grazing-incidence Bragg-case geometry. *Philosophical Magazine* 92.36 (2012): 4599–4617.
36. Fundenberger, J. J., Morawiec, A., Bouzy, E., and Lecomte, J. S. Polycrystal orientation maps from TEM. *Ultramicroscopy* 96.2 (2003): 127–137.
37. Zandiatashbar, A., Lee, G. H., An, S. J., Lee, S., Mathew, N., Terrones, M., and Koratkar, N. Effect of defects on the intrinsic strength and stiffness of graphene. *Nature Communications* 5.1 (2014): 3186.
38. Wang, X., Wu, J., Zhang, Y., Sun, Y., Ma, K., Xie, Y., and Zhang, Y. Vacancy defects in 2D transition metal dichalcogenide electrocatalysts: From aggregated to atomic configuration. *Advanced Materials* 35.50 (2023): 2206576.
39. Santos, A. J., Lacroix, B., Maudet, F., Paumier, F., Hurand, S., Dupeyrat, C., and Morales, F. M. Application of advanced (S) TEM methods for the study of nanostructured porous functional surfaces: A few working examples. *Materials Characterization* 185 (2022): 111741.
40. Baykara, M. Z., and Schwarz, U. D. Atomic force microscopy: Methods and applications. In *Encyclopedia of spectroscopy and spectrometry*. Elsevier Science, 2017: 70–75.
41. Taherimakhsousi, N., MacLeod, B. P., Parlane, F. G., Morrissey, T. D., Booker, E. P., Dettelbach, K. E., and Berlinguette, C. P. Quantifying defects in thin films using machine vision. *npj Computational Materials* 6.1 (2020): 111.
42. Alsteens, D., Gaub, H. E., Newton, R., Pfreundschuh, M., Gerber, C., and Müller, D. J. Atomic force microscopy-based characterization and design of biointerfaces. *Nature Reviews Materials* 2.5 (2017): 1–16.
43. Barth, C., Foster, A. S., Henry, C. R., and Shluger, A. L. Recent trends in surface characterization and chemistry with high-resolution scanning force methods. *Advanced Materials* 23.4 (2011): 477–501.
44. Kaupp, G. *Atomic force microscopy, scanning nearfield optical microscopy and nanoscratching: Application to rough and natural surfaces*. Springer Science & Business Media, 2006.
45. Zhang, F., Edwards, D., Deng, X., Wang, Y., Kilpatrick, J. I., Bassiri-Gharb, N., and Rodriguez, B. J. Investigation of AFM-based machining of ferroelectric thin films at the nanoscale. *Journal of Applied Physics* 127.3 (2020).
46. Passeri, D., Rossi, M., Tamburri, E., and Terranova, M. L. Mechanical characterization of polymeric thin films by atomic force microscopy based techniques. *Analytical and Bioanalytical Chemistry* 405 (2013): 1463–1478.
47. Fricker, A. L., McPhail, D. S., Keneghan, B., and Pretzel, B. Investigating the impact of cleaning treatments on polystyrene using SEM, AFM and ToF–SIMS. *Heritage Science* 5 (2017): 1–9.

48. Rao, M. C., and Shekhawat, M. S. A brief survey on basic properties of thin films for device application. *International Journal of Modern Physics: Conference Series* 22 (2013): 576–582.
49. Abbas, W., Ibrahim, M. S., Waseem, M., Lu, C., Lee, H. H., Fazal, S., and Pramanick, A. defect and texture engineering of relaxor thin films for High-Power energy storage applications. *Chemical Engineering Journal* (2024): 148943.
50. Yao, W., Wu, B., and Liu, Y. Growth and grain boundaries in 2D materials. *ACS Nano* 14.8 (2020): 9320–9346.
51. Thompson, C. V. Solid-state dewetting of thin films. *Annual Review of Materials Research* 42 (2012): 399–434.
52. Zhou, Q., Ren, Y., Du, Y., Hua, D., and Han, W. Cracking and toughening mechanisms in nanoscale metallic multilayer films: A brief review. *Applied Sciences* 8.10 (2018): 1821.
53. Anderson, C. V., and Tamma, K. K. An overview of advances in heat conduction models and approaches for prediction of thermal conductivity in thin dielectric films. *International Journal of Numerical Methods for Heat & Fluid Flow* 14.1 (2004): 12–65.
54. Morosanu, C. E. *Thin films by chemical vapour deposition*. Elsevier, 2016.
55. Phiri, R. R., Oladijo, O. P., and Akinlabi, E. T. defect formation and surface evolution of thin film materials. *Proceedings of the International Conference on Industrial Engineering and Operations Management* (2018): 1087–1093.
56. Kapoor, M., and Thompson, G. B. Role of atomic migration in nanocrystalline stability: Grain size and thin film stress states. *Current Opinion in Solid State and Materials Science* 19.2 (2015): 138–146.
57. Abate, A. A., Piqueras, C. M., and Vega, D. A. Defect-induced order–order phase transition in triblock copolymer thin films. *Macromolecules* 54.4 (2021): 1639–1647.
58. Kumar, R., Kumar, G., Al-Dossary, O., and Umar, A. ZnO nanostructured thin films: Depositions, properties and applications – A review. *Materials Express* 5.1 (2015): 3–23.
59. Yan, X., Wei, J., An, K., Liu, J., Chen, L., Zheng, Y., and Li, C. High temperature surface graphitization of CVD diamond films and analysis of the kinetics mechanism. *Diamond and Related Materials* 120 (2021): 108647.
60. Evans, A. G., and Hutchinson, J. W. The thermomechanical integrity of thin films and multilayers. *Acta Metallurgica et Materialia* 43.7 (1995): 2507–2530.
61. Shugurov, A. R., and Panin, A. V. Mechanisms of stress generation in thin films and coatings. *Technical Physics* 65 (2020): 1881–1904.
62. Innocenzi, P., and Malfatti, L. Mesoporous thin films: Properties and applications. *Chemical Society Reviews* 42.9 (2013): 4198–4216.
63. Opila, R. L., and Eng Jr, J. Thin films and interfaces in microelectronics: Composition and chemistry as function of depth. *Progress in Surface Science* 69.4–6 (2002): 125–163.
64. Setter, N., Damjanovic, D., Eng, L., Fox, G., Gevorgian, S., Hong, S., and Streiffer, S. Ferroelectric thin films: Review of materials, properties, and applications. *Journal of Applied Physics* 100.5 (2006).
65. Li, M., Zhu, Z., Wang, Z., Pan, W., Cao, X., Wu, G., and Chen, R. High-quality hybrid perovskite thin films by post-treatment technologies in photovoltaic applications. *Advanced Materials* 36.7 (2024): 2309428.
66. Moreno, M., Patriarche, G., and i Cabarrocas, P. R. Fine-tuning of the interface in high-quality epitaxial silicon films deposited by plasma-enhanced chemical vapor deposition at 200 C. *Journal of Materials Research* 28.13 (2013): 1626–1632.
67. Chambers, S. A. Epitaxial growth and properties of thin film oxides. *Surface Science Reports* 39.5–6 (2000): 105–180.
68. Kashyap, A., Singh, N. K., Soni, M., and Soni, A. deposition of thin films by chemical solution-assisted techniques. In *Chemical solution synthesis for materials design and thin film device applications*. Elsevier, 2021: 79–117.

69. Mattox, D. M. *Handbook of physical vapor deposition (PVD) processing*. William Andrew, 2010.
70. Nehm, F., Klumbies, H., Richter, C., Singh, A., Schroeder, U., Mikolajick, T., and Müller-Meskamp, L. Breakdown and protection of ALD moisture barrier thin films. *ACS Applied Materials & Interfaces* 7.40 (2015): 22121–22127.
71. Todorov, T., Hillhouse, H. W., Aazou, S., Sekkat, Z., Vigil-Galán, O., Deshmukh, S. D., and Dale, P. J. Solution-based synthesis of kesterite thin film semiconductors. *Journal of Physics: Energy* 2.1 (2020): 012003.
72. Liu, J., and Wöll, C. Surface-supported metal–organic framework thin films: Fabrication methods, applications, and challenges. *Chemical Society Reviews* 46.19 (2017): 5730–5770.
73. Acosta, E. Thin films/properties and applications. In *Thin films*. IntechOpen, 2021.
74. Paz, Y. Self-assembled monolayers and titanium dioxide: From surface patterning to potential applications. *Beilstein Journal of Nanotechnology* 2.1 (2011): 845–861.
75. Ghosh, R., Paul, G. K., and Basak, D. Effect of thermal annealing treatment on structural, electrical and optical properties of transparent sol–gel ZnO thin films. *Materials Research Bulletin* 40.11 (2005): 1905–1914.
76. Priebe, A., Dousse, B., Tzou, C. Y., Papadopoulos, G., Utke, I., Bensaoula, A., and Guerra-Nuñez, C. Real-time in situ parallel detection of elements and molecules with TOFMS during ALD for chemical quality control of thin films. *The Journal of Physical Chemistry C* 126.4 (2022): 1901–1912.
77. Haider, A. J., Alawsi, T., Haider, M. J., Taha, B. A., and Marhoon, H. A. A comprehensive review on pulsed laser deposition technique to effective nanostructure production: Trends and challenges. *Optical and Quantum Electronics* 54.8 (2022): 488.
78. Brabazon, D., and Raffer, A. Advanced characterization techniques for nanostructures. In *Emerging nanotechnologies for manufacturing*. Elsevier, 2015: 53–85.
79. Chen, X., Cheng, S., Xiao, L., and Sun, H. Identifying, understanding and controlling defects and traps in halide perovskites for optoelectronic devices: A review. *Journal of Physics D: Applied Physics* 53.37 (2020): 373001.
80. Franklin, A. D. Nanomaterials in transistors: From high-performance to thin-film applications. *Science* 349.6249 (2015): aab2750.
81. Khan, K., Tareen, A. K., Aslam, M., Wang, R., Zhang, Y., Mahmood, A., and Guo, Z. Recent developments in emerging two-dimensional materials and their applications. *Journal of Materials Chemistry C* 8.2 (2020): 387–440.
82. Butt, M. A., Tyszkiewicz, C., Karasiński, P., Zięba, M., Kaźmierczak, A., Zdończyk, M., and Piramidowicz, R. Optical thin films fabrication techniques – Towards a low-cost solution for the integrated photonic platform: A review of the current status. *Materials* 15.13 (2022): 4591.
83. Myny, K. The development of flexible integrated circuits based on thin-film transistors. *Nature Electronics* 1.1 (2018): 30–39.
84. Alagappan, M., Sridarshini, T., and Sowbarnika, P. S. Flexible thin films for electronics. In *Advances in flexible and printed electronics: Materials, fabrication, and applications*. IOP Publishing, 2023: 3–1.
85. Toma, F. T. Z., Rahman, S., Hussain, K. M., and Ahmed, S. Thin film deposition techniques: A comprehensive review. *Journal of Modern Nanotechnology* 4.6 (2024).
86. Ng, L. W., Lee, S. W., Chang, D. W., Hodgkiss, J. M., and Vak, D. Organic photovoltaics' new renaissance: Advances toward roll-to-roll manufacturing of non-fullerene acceptor organic photovoltaics. *Advanced Materials Technologies* 7.10 (2022): 2101556.
87. Abdulraheem, M. I., Moshood, A. Y., Chen, Y., Chen, H., Zhang, H., and Hu, J. Advancements in designing smart and intelligent nanocoatings. In *Sustainable approach to protective nanocoatings*. IGI Global Publisher, 2024: 57–87.

88. Montemor, M. F. Functional and smart coatings for corrosion protection: A review of recent advances. *Surface and Coatings Technology* 258 (2014): 17–37.
89. Zhang, X., Chen, L., Lim, K. H., Gonuguntla, S., Lim, K. W., Pranantyo, D., and Soh, S. The pathway to intelligence: Using stimuli-responsive materials as building blocks for constructing smart and functional systems. *Advanced Materials* 31.11 (2019): 1804540.
90. Horne, J., McLoughlin, L., Bury, E., Koh, A. S., and Wujcik, E. K. Interfacial phenomena of advanced composite materials toward wearable platforms for biological and environmental monitoring sensors, armor, and soft robotics. *Advanced Materials Interfaces* 7.4 (2020): 1901851.
91. Chao, H. Y., Venkatraman, K., Moniri, S., Jiang, Y., Tang, X., Dai, S., and Chi, M. In situ and emerging transmission electron microscopy for catalysis research. *Chemical Reviews* 123.13 (2023): 8347–8394.
92. Atkins, D., Capria, E., Edström, K., Famprikis, T., Grimaud, A., Jacquet, Q., and Lyonnard, S. Accelerating battery characterization using neutron and synchrotron techniques: Toward a multi-modal and multi-scale standardized experimental workflow. *Advanced Energy Materials* 12.17 (2022): 2102694.
93. Kunde, G. B., and Sehgal, B. Industrial innovation through sustainable materials. In *Handbook of smart materials, technologies, and devices: Applications of Industry 4.0.* Springer, 2021: 1–42.
94. Lejarazu-Larrañaga, A., Landaburu-Aguirre, J., Senán-Salinas, J., Ortiz, J. M., and Molina, S. Thin film composite polyamide reverse osmosis membrane technology towards a circular economy. *Membranes* 12.9 (2022): 864.
95. Allan, J. E., Ferreri, A., Grande, S., Giannantonio, R., and Matteucci, F. *Technology transfer in nanotechnology*. Publications Office of the European Union, Luxembourg, 2019.
96. Natarajan, S. *Nanotechnology for the energy challenge*, edited by Javier Garcia-Martinez. Wiley-VCH Publishing Series, 2013, 639 pp. ISBN: 978-3-527-33380-6 (PRINT). ISBN: 978-3-527-66513-6 (ePDF).
97. Seshan, K., and Schepis, D. (Eds.). *Handbook of thin film deposition.* William Andrew, 2018.
98. Vijayan, K., Vijayachamundeeswari, S. P., Sivaperuman, K., Ahsan, N., Logu, T., and Okada, Y. A review on advancements, challenges, and prospective of copper and non-copper based thin-film solar cells using facile spray pyrolysis technique. *Solar Energy* 234 (2022): 81–102.

6 Hydrophobicity and Hydrophilicity in Thin Films

Nature's Inspiration to Application

Yadav Narendra Kumar Rajbahadur, Alfa Bisoi, and Simanchal Kar

6.1 INTRODUCTION

The wetting phenomenon on a surface is a critical factor in determining how fluids interact with solids. The degree to which a liquid adheres to a solid surface is governed by the comparative strength of the forces that bind the liquid to the solid versus those that hold the liquid together [1, 2]. This interaction is quantified using a force balance equation that assesses the contact angle created by a droplet on the surface, which serves as an indicator of the surface's wetting potential. The contact angle (θ) is the angle formed where a liquid, a solid, and a gas (or vapour) meet. Specifically, it's the angle between the tangent line to the liquid's surface and the solid surface, measured through the liquid. A higher contact angle indicates a better hydrophobic (water-repellent) surface, where the fluid has minimal contact with the solid. Conversely, a lower contact angle suggests a more hydrophilic (water-attractive) surface, where the fluid spreads out more extensively. Thus, for analysing the wettability of a surface, the contact angle is a crucial parameter. This crucial surface property, based on the wettability, has encouraged several applications like self-cleaning surfaces, coatings, and a few biocompatible materials [2–5]. In essence, it is a fundamental parameter for tailoring surface properties for various applications. Figure 6.1 shows the presence of hydrophobic and hydrophilic surface in nature.

The presence of hydrophobicity was first observed in living organisms to cope with nature. For example, duck feather, insect cuticle, wings of butterfly, fish scale, and water strider legs have minimum surface energy to reduce wetting [8–11]. Superhydrophobic surfaces, exhibiting exceptional water-repelling abilities, are gaining significant attention due to their potential in self-cleaning, anti-fouling, water resistance, and anti-stick applications. These materials, inspired by nature, are defined by a water contact angle greater than 150° and a sliding angle less than 10° [12–14]. Figure 6.2 illustrates the different contact angles observed under various wetting conditions.

DOI: 10.1201/9781003750659-6

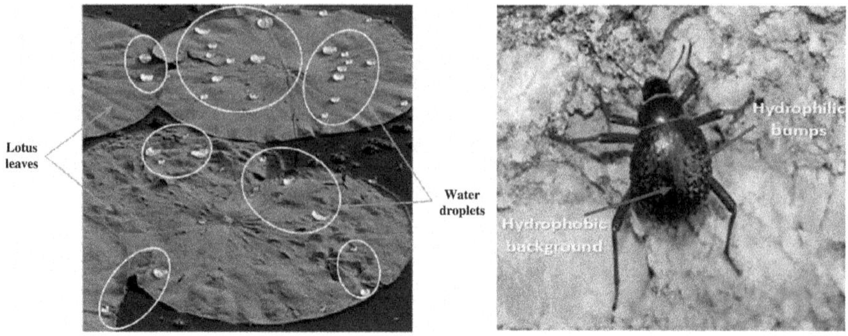

FIGURE 6.1 Presence of hydrophobicity and hydrophilicity in nature [6, 7].

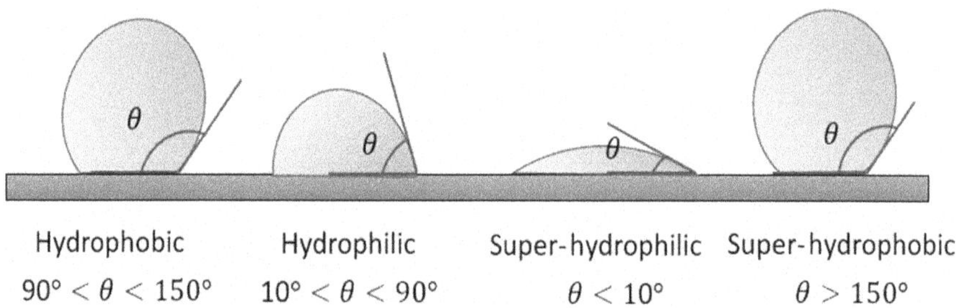

FIGURE 6.2 Different types of surfaces depending on the contact angle [15].

Hydrophilic surfaces, inspired by nature, exhibit remarkable wetting properties that enable them to attract and spread water effectively. One of the most notable examples of natural hydrophilic surfaces is the structure of plant leaves, such as those of the camellia plant, which have evolved to spread water evenly, enhancing water absorption [16, 17]. These surfaces possess micro- and nanostructures that enhance water adhesion, allowing for efficient water distribution and minimizing water loss. Additionally, certain insects and amphibians display hydrophilic skin that aids in hydration and thermoregulation [18, 19]. Mimicking these natural designs, scientists and engineers have developed synthetic hydrophilic surfaces for various applications, including medical implants that promote cell adhesion and tissue integration, coatings that improve the efficacy of water-based paints and adhesives, and materials that enhance the performance of desalination membranes and filtration systems [5, 20, 21].

The hydrophobic or hydrophilic nature of a surface is largely affected by asperities and surface energy. Surface roughness amplifies the inherent wetting properties of a surface. Rough surfaces can trap air pockets, increasing hydrophobicity by preventing water from fully spreading, as seen in the lotus leaf effect. Conversely, roughness can enhance hydrophilicity by providing more surface area for water to adhere, facilitating water spreading [2, 22, 23].

Surface energy, the measure of surplus energy at a material's surface, also plays a crucial role. High surface energy materials attract water molecules, promoting hydrophilicity and resulting in lower contact angles. In contrast, low surface energy materials repel water, leading to higher contact angles and increased hydrophobicity. The interplay between surface roughness and surface energy allows for precise control over a material's wetting properties, enabling the design of surfaces with specific hydrophobic or hydrophilic characteristics for various applications [5, 21, 24, 25].

6.2 UNDERSTANDING WETTABILITY: WETTING MODELS THAT GOVERN HYDROPHOBICITY AND HYDROPHILICITY

6.2.1 Young's Model

Young's wetting model describes the equilibrium contact angle of a liquid droplet on a solid surface, and it gives details of wetting behaviour of materials. According to Thomas Young's theory, the contact angle (θ_{eq}) is described as the equilibrium of interfacial tensions between the solid, liquid, and vapour phases [26–28]. The Young equation is given by Equation 1, and Young's model is shown in Figure 6.2 (a).

$$\cos\theta_{eq} = \frac{\gamma_{sv} - \gamma_{sl}}{\gamma_{lv}} \tag{1}$$

Where γ_{lv}, γ_{sv}, and γ_{sl} are liquid–vapour interfacial tension, solid–vapour interfacial tension, solid–liquid interfacial tension, respectively.

In this model, the contact angle θ quantifies the wetting angle of a surface. A low contact angle ($\theta < 90°$) indicates good wetting and hydrophilicity, meaning the liquid evenly spreads out on the surface. A greater contact angle ($\theta > 90°$) signifies less wetting and hydrophobicity, where the liquid forms a droplet and minimizes contact with the surface. A perfect, smooth, and chemically homogenous surface is assumed by Young's model. However, real-world surfaces often exhibit roughness and chemical heterogeneity, leading to deviations from the ideal predictions. Despite these limitations, Young's wetting model serves as a foundational concept in surface science, helping to understand and predict the wetting behaviour of several materials, and it leads to the design of surfaces with tailored wettability for various applications [27, 29, 30].

6.2.2 Wenzel's Model

Wenzel's wetting model, proposed by Robert N. Wenzel in 1936, provides a framework for understanding how surface roughness influences the wettability of a material [31]. In this model, the apparent contact angle (θ^*) on a rough surface is given by Equation 2 suggested in Wenzel model. (see Figure 6.2(b)).

$$\cos\theta^* = r\cos\theta \tag{2}$$

where θ indicates the contact angle on a smooth surface, and 'r' is the roughness factor, calculated as the actual surface area divided by the projected surface area.

Wenzel's model states that because of the increased surface area, the apparent contact angle between a liquid droplet and a rough surface is dissimilar from that on a smooth surface. For inherently hydrophobic surfaces (where θ is greater than $90°$), higher roughness (expressed as the roughness factor *r*) amplifies the apparent contact angle, enhancing hydrophobicity. Conversely, for inherently hydrophilic surfaces (less than $90°$), increased roughness reduces the apparent contact angle, thereby boosting hydrophilicity. Wenzel's model assumes complete wetting of the surface by the liquid, filling in all surface irregularities. This understanding is pivotal for designing surfaces with specific wetting properties tailored for applications ranging from coatings and biomedical devices to self-cleaning materials, providing a foundational framework for advancing surface engineering and material science [29, 32, 33].

6.2.3 Cassie-Baxter Model

The Cassie-Baxter wetting model is a theoretical framework that describes the wetting condition of a liquid droplet on a rough or textured surface. Named after the scientists who proposed it in the 1940s, W. H. Cassie and S. Baxter, this model contrasts with the simpler Wenzel model by considering how air pockets trapped within surface asperities affect wetting. This leads to a composite interface where the apparent contact angle (θ_c) is distinct from the intrinsic contact angle (θ) predicted by the Wenzel model [34, 35]. This model is shown in Figure 6.2(c) and represented by Equation 3.

$$\cos\theta_c = f\cos\theta + (1 - f) \tag{3}$$

The intrinsic contact angle θ defines the liquid's behaviour on a perfectly flat surface, while 'f' quantifies the actual area of solid-liquid contact.

The Cassie-Baxter state is characterized by a liquid droplet that is suspended on the tips of surface protrusions, with air filling the spaces in between. This arrangement lowers the solid–liquid contact area compared to a flat surface, leading to an increased apparent contact angle. Therefore, the surface appears more hydrophobic or hydrophilic depending on its intrinsic properties and the nature of the liquid [32, 35, 36].

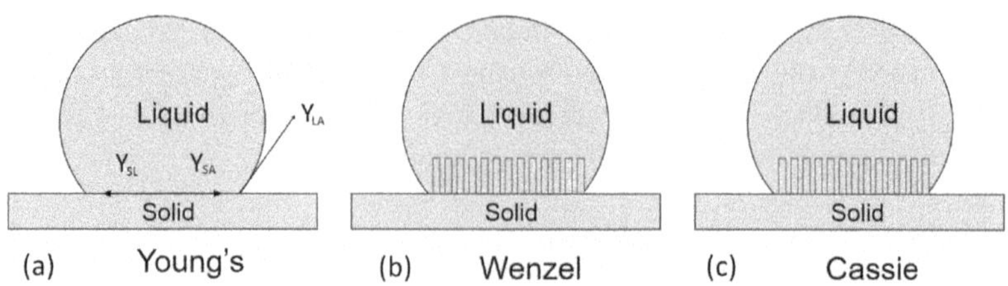

FIGURE 6.3 Wetting models (a) Young, (b) Wenzel, and (c) Cassie-Baxter wetting models [37].

6.3 METHODS FOR APPLYING THIN FILMS TO CREATE HYDROPHOBIC OR HYDROPHILIC SURFACES

6.3.1 Dip Coating

Dip coating method is used for applying a uniform coating on every side of a material's surface. It is highly effective for applications needing thin coatings, such as surface painting. The technique involves several steps: immersion, deposition, evaporation, and drainage [38]. The film's thickness can be changed by adjusting the withdrawal speed, which depends on the viscosity and density of the sol-gel [39]. This method excels in quickly coating both sides of a material prior to production. Because of its scalability, it provides an affordable option for operations involving mass production. The technique's broad acceptance can be credited to its ease of use and suitability for a variety of coating materials. Compared to alternative deposition techniques, cost-effectiveness is the consequence of little waste, effective material usage, and decreased energy consumption [38, 40, 41]. For example, Shing et al. [42] employed dip coating to develop a thin film coating on a glass panel using silica and hexadimethylsilane. The coating exhibited remarkable super hydrophobicity of 161° and increased the transmission of glass by 5%. Figure 6.4 shows the schematic diagram of the steps involved in the dip coating method.

6.3.2 Spin Coating

The spin coating process involves dispersing a solution uniformly onto a surface by spinning, using heat to ensure adhesion. The thin film's thickness is adjusted through changes in rotational velocity, spin time, and solution viscosity. This method generally involves applying a sol-gel solution, followed by controlled spinning to create a thin film, and then drying through evaporation [43, 44]. Spin coating is widely favoured for its simplicity, speed, and cost-effectiveness in depositing thin,

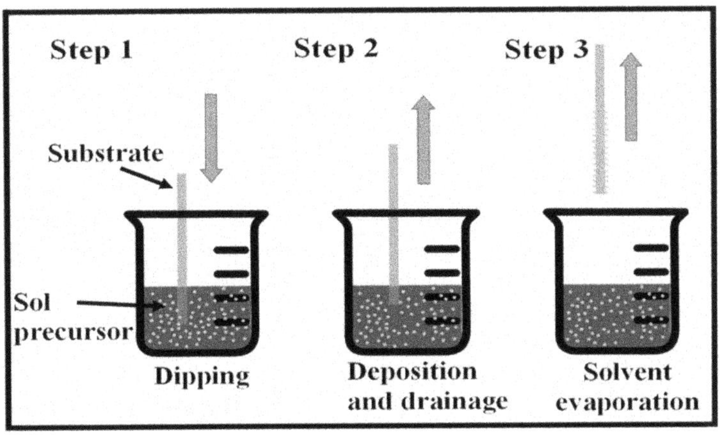

FIGURE 6.4 Different steps of dip coating [2].

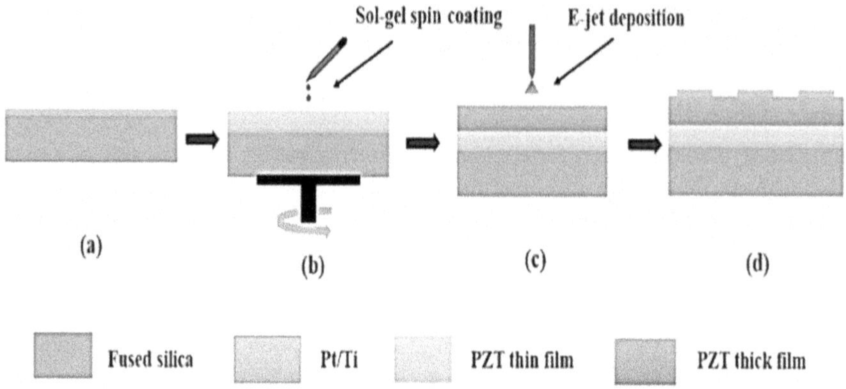

FIGURE 6.5 Basic steps for thin film coating of different layers using spin coating [48].

uniform films, adaptable to various substrates like glass and silicon. While suitable for research and development due to low-temperature processing, challenges include achieving perfect uniformity, edge exclusion effects, and limitations with thick films [45]. Other factors to take into account are material waste, solvent sensitivity, and material compatibility. Thorough optimization is necessary when increasing production and covering intricate or three-dimensional structures [46]. For example, Sutha et al. [47] created a super-hydrophobic thin film on a glass surface using spin coating and heating each layer to 150 °C for 1 minute to increase adhesion and then annealing the sample at 400 °C for 1 hour. This coated sample resulted in a contact angle of 161°. Figure 6.5 illustrates the spin coating application process.

6.3.3 Spray Coating

Spray pyrolysis process is a flexible and effective method for applying thin films on a range of surfaces. Using this technique, the intended coating materials are created as a fine mist or spray, which is then subjected to heat. A thin, homogenous covering is formed as a result of the sprayed solution's thermal breakdown [40, 49].

This method has a number of benefits. It is ideal for a variety of industrial applications due to its high versatility and scalability, which enable the formation of thin films across huge surface and complex geometries. Furthermore, spray pyrolysis is affordable, offering a practical way to cover surfaces with different materials. There are restrictions, though. It might be difficult to get exact control over layer thickness and homogeneity, requiring careful optimization of deposition parameters [50].

Spraying droplets during the process and requiring high-temperature processing can limit the application of this technique on materials or substrates that are heat-sensitive. Despite these difficulties, spray pyrolysis coating is a useful technique for thin film formation in a variety of real-world scenarios due to its scalability and affordability [51]. For instance, Tarwal et al. [52] employed spray pyrolysis to form a transparent, superhydrophobic ZnO thin layer on a glass substrate. With a water contact angle of 154°, the coated substrate had an impressive surface that gave it

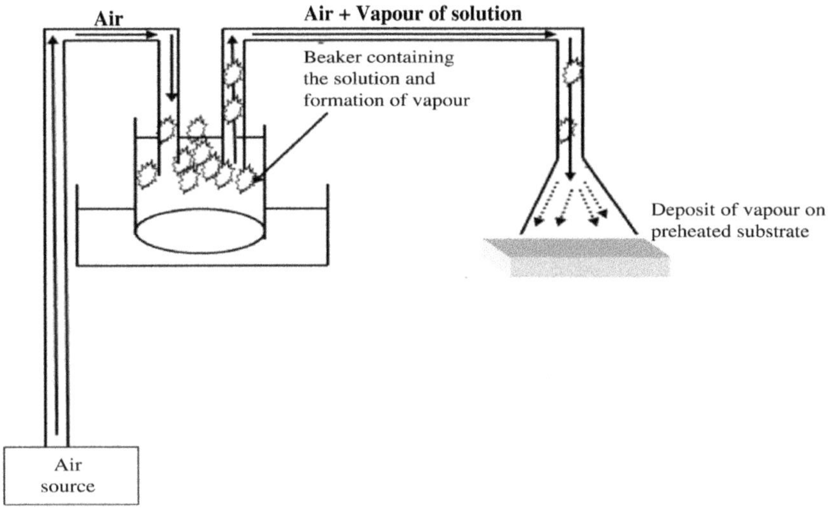

FIGURE 6.6 Schematic of spray pyrolysis deposition technique [53].

extraordinary hydrophobicity. The spray pyrolysis method's schematic for film deposition is displayed in Figure 6.6.

6.3.4 Chemical Vapour Deposition

Among the popular coating methods, chemical vapour deposition (CVD) stands out for its versatility and precision, like sol-based methods. CVD excels in depositing uniform coatings of various materials on complex surfaces, providing excellent controllability over the structure and thickness of the film, which makes it indispensable for materials research, nanotechnology, and microelectronics. Its gas-phase reactions enable the production of high-purity materials, and its conformal layering ability allows for modification of material properties, benefitting many industries [54].

Nevertheless, there are drawbacks to CVD, such as the requirement for suitable precursors, high temperatures for specific operations, and expensive and complicated equipment. Slow deposition rates and by-product generation also pose challenges. The current study aims to increase CVD's efficacy and address these drawbacks to maintain its relevance in advanced technologies and large-scale manufacturing. For example, Young et al. [55] developed an amorphous thin film of titanium dioxide for heat exchanger using CVD. The film exhibited water contact angle of 8°, indicating super hydrophilic nature with enhanced durability [54, 56, 57]. A schematic of the CVD method's film deposition is represented in Figure 6.7.

6.3.5 RF Magnetron Sputtering

RF magnetron sputtering is a cutting-edge process for depositing thin films. It utilizes radiofrequency energy to generate a dense plasma within a vacuum, causing

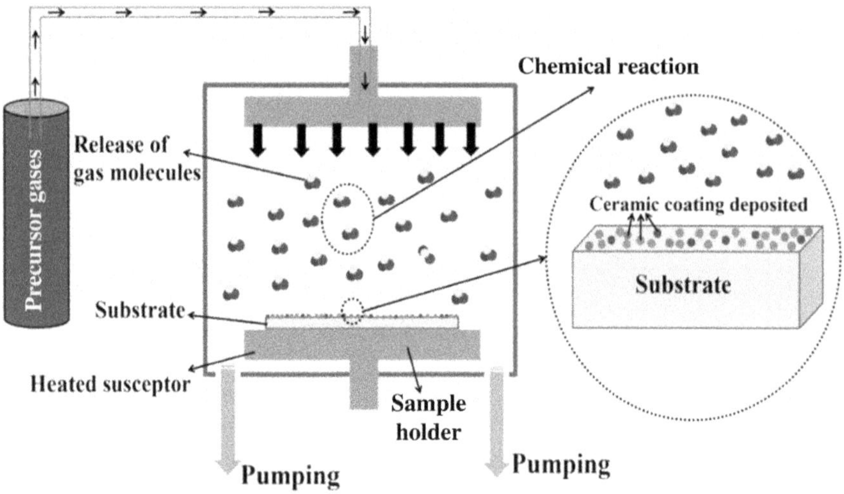

FIGURE 6.7 Schematic of CVD [58].

atoms from a target material to be ejected. These energetic particles then condense onto a substrate, resulting in a precise and even coating [59–61].

This process offers multiple benefits, including strong substrate adhesion, rapid deposition speeds, precise control over coating thickness, and uniform film deposition across large and complex surfaces. Its adaptability and extensive application are further enhanced by its ability to function at low temperatures and accommodate a variety of materials. However, RF magnetron sputtering has limitations. The equipment is complex and costly, target utilization efficiency is low, resulting in higher operational costs, and to achieve uniform coating 3D surfaces can be tricky. There are also risks of cross-contamination and high energy consumption, which have an effect on expenses and environmental factors. Despite these drawbacks, RF magnetron sputtering remains an essential technique in thin-film deposition, used in functions ranging from optical coatings to semiconductor devices [61, 62]. For example, Liu et al. [63] fabricated a thin film composite of SiO_2/TiO_2. The coating resulted in super hydrophilicity having a water contact angle of 0° for glass when annealed at 500 °C. Figure 6.8 represents the schematic diagram of RF magnetron sputtering process.

6.4 APPLICATIONS OF HYDROPHOBIC SURFACES

Hydrophobic surfaces are employed over a wide range of industries due to their water-repellent properties. In the automotive industry, they are used on windshields, mirrors, and car bodies to enhance visibility and reduce maintenance by preventing water, mud, and other contaminants from adhering to surfaces. This results in improved safety and a cleaner appearance. In electronics, hydrophobic coatings protect sensitive components from moisture, which can cause short circuits and corrosion, thereby extending the lifespan and reliability of devices such as smartphones,

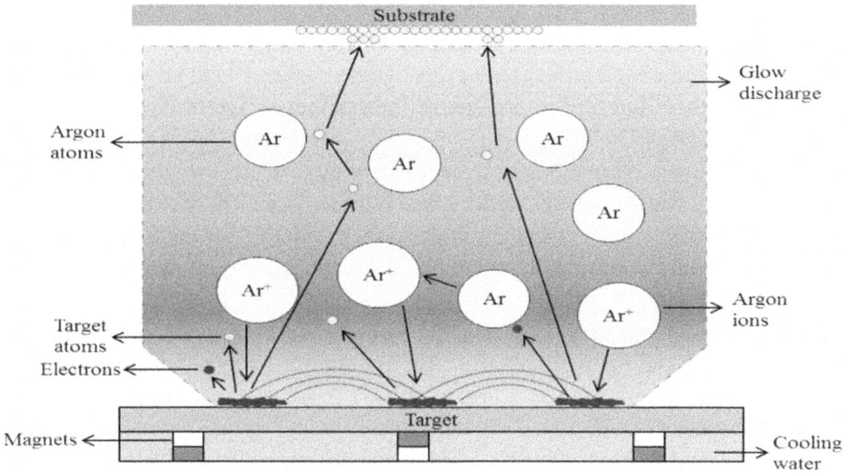

FIGURE 6.8 Schematic illustration of RF magnetron sputtering process [64].

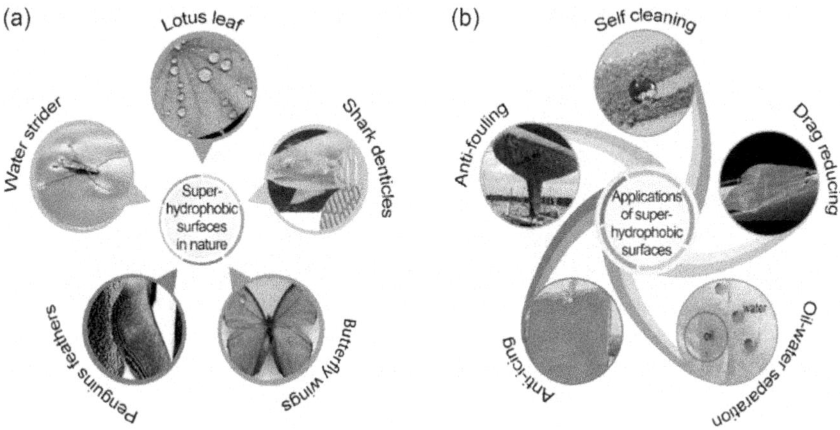

FIGURE 6.9 (a) Hydrophobicity in nature, and (b) Practical application of hydrophobic surfaces [63].

tablets, and wearables [65–68]. Figure 6.9 shows the presence of hydrophobicity in nature and its various applications in the real world.

Medical field: Hydrophobic surfaces are applied to surgical instruments, medical devices, and implants to reduce the risk of contamination and improve hygiene. These coatings help prevent bacterial adhesion and biofilm formation, which are critical for maintaining sterile conditions and reducing infection rates. Hydrophobic materials are also used in developing water-resistant dressings and wound care products, which keep wounds dry and promote faster healing [69–71].

Construction industry: Hydrophobic coatings are applied to building materials like concrete, glass, and wood to protect them from water damage, thus increasing durability and reducing maintenance costs. These surfaces help prevent issues such as mould growth, efflorescence, and freeze-thaw damage, thereby enhancing the longevity of structures [72–74].

Textiles industry: This industry benefits from hydrophobic treatments applied to fabrics, creating water-resistant clothing and outdoor gear. These treatments provide dryness and comfort in wet weather, making them suitable for outdoor pursuits and athletic endeavours. Additionally, hydrophobic coatings are used in footwear to keep shoes dry and clean [75, 76].

Food packaging industry: Hydrophobic surfaces are used to create non-stick and easy-to-clean packaging materials that help maintain food hygiene and extend shelf life. These coatings prevent the adhesion of food residues, making packaging easier to clean and reuse [77, 78].

Aerospace and marine industries: This industry also utilizes hydrophobic coatings to protect aircraft and ships from corrosion, icing, and fouling. In aerospace, these coatings help maintain the integrity of critical components exposed to harsh environmental conditions, while in marine applications, they reduce drag and prevent biofouling, enhancing fuel efficiency and performance [71, 79, 80]. Overall, the versatility and effectiveness of hydrophobic surfaces make them indispensable in numerous practical applications across different fields, contributing to improved performance, durability, and hygiene.

6.5 APPLICATIONS OF HYDROPHILIC SURFACES

The ability of hydrophilic surfaces to engage with water molecules makes them valuable in many different industries. In fog and dew harvesting systems, hydrophilic surfaces play a crucial role in capturing water from fog and dew. Inspired by nature, these surfaces promote condensation, allowing droplets to form and collect. In regions with limited freshwater resources, such systems provide an eco-friendly way to supplement water supply [81, 82]. Figure 6.10 shows the presence of hydrophilicity in nature and its various real-world applications.

- Moisture Harvesting in Arid Environments: Drawing inspiration from natural adaptations like the Namib desert beetle, researchers have developed hydrophilic materials capable of harvesting moisture from the atmosphere. These innovative materials are applicable in various fields such as greenhouses, agricultural techniques, and emergency survival kits. By emulating the microstructures of the beetle, these surfaces efficiently absorb water, offering a sustainable solution to support communities in arid environments and challenging climates [81, 86, 87].
- Solar Desalination: Solar-driven desalination relies on hydrophilic membranes. These membranes allow water vapour to pass through while blocking salt ions. As sunlight heats seawater, the hydrophilic surface ensures effective vapour transport. This results in clean, desalinated water ready for consumption. This technology holds promise for coastal regions facing water scarcity [88, 89].

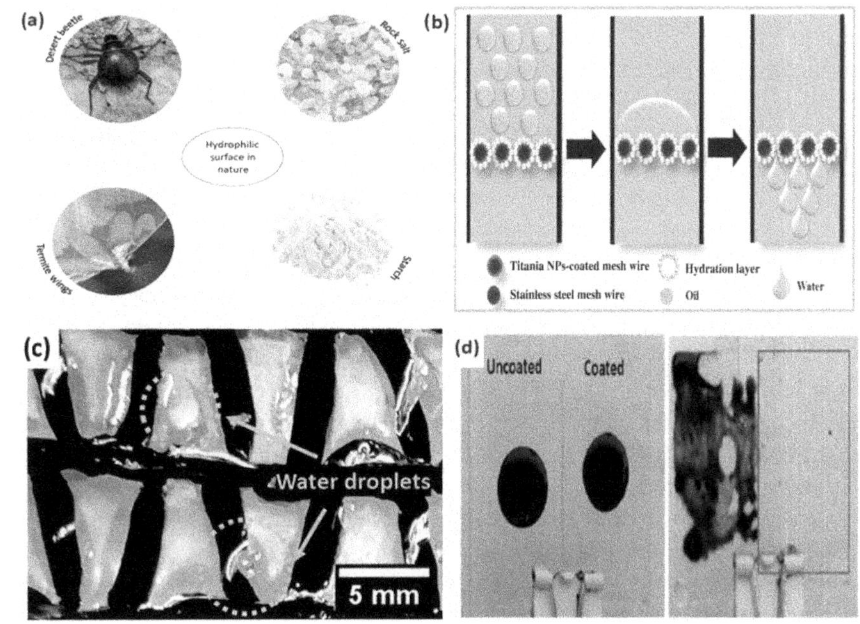

FIGURE 6.10 (a) Hydrophilicity in nature, (b) Oil–water separation through hydrophobic membrane [83], (c) Fog collection through hydrophilic coated raschel mesh [84], and (d) Self-cleaning of glass through hydrophilic coating [85].

- Biomedical Devices: Hydrophilic coatings are crucial for medical devices such as catheters, stents, and implants. These coatings reduce friction, minimize bacterial adhesion, and enhance biocompatibility, leading to improved patient outcomes and reduced risk of infection. Additionally, hydrophilic coatings on diagnostic equipment improve the accuracy and reliability of tests by preventing sample contamination and ensuring consistent fluid flow [90, 91].
- Enhanced Cooling Systems: Heat exchangers, radiators, and air-conditioning units benefit from hydrophilic surfaces. By improving cooling efficiency, these materials ensure optimal performance. Heat transfer becomes more effective, whether in industrial processes or everyday appliances [92, 93].
- Lab-on-a-Chip Technology: Microfluidic devices, often used in diagnostics and DNA analysis, rely on hydrophilic channels. These tiny chips manipulate fluids precisely, enabling rapid tests and research. Whether detecting diseases or studying cellular interactions, hydrophilic surfaces ensure controlled flow and efficient reactions [94–96].

In summary, hydrophilic surfaces offer practical solutions across disciplines, from water conservation to advanced technologies. By learning from nature, we create materials that enhance our lives and protect our planet.

6.6 FUTURE WORK

Future advancements in hydrophobic and hydrophilic coatings offer substantial potential for diverse applications. These coatings continue to be pivotal in various industries, with ongoing research aimed at advancing their capabilities and applications. For hydrophilic coatings, current efforts focus on optimizing surface modification techniques and material formulations to enhance water affinity and durability. Researchers are exploring innovations in biomedical applications, aiming to improve the biocompatibility of medical implants and devices while also mitigating bacterial adhesion through advanced coating technologies. Conversely, in the realm of hydrophobic coatings, the emphasis lies on improving durability and expanding functionalities beyond mere water repellence. Novel approaches involve the development of robust nanostructures and composite materials that offer superior wear resistance and self-cleaning properties. Applications in automotive and aerospace sectors are exploring coatings that resist corrosion, reduce drag, and enhance fuel efficiency and overall performance.

Interdisciplinary research is also exploring the integration of smart materials and responsive surfaces into both hydrophilic and hydrophobic coatings. These innovations aim to create coatings capable of dynamically adjusting their properties in response to environmental changes, such as temperature and humidity. Such adaptive coatings hold promise for applications in wearable technology, environmental sensing, and beyond, where responsiveness to changing conditions can significantly enhance functionality and longevity.

In summary, ongoing research in hydrophilic and hydrophobic coatings is advancing the frontiers of material science, paving the way for new and enhanced applications across diverse industries.

REFERENCES

1. Novikova A, Katiyi A, Karabchevsky A. Nature-inspired anti-reflective texturization for solar energy applications. *Adv Mater Technol*. 2023 Nov 27;9(2).
2. Butt MA. Thin-film coating methods: A successful marriage of high-quality and cost-effectiveness – a brief exploration. *Coatings*. 2022 Aug 4;12(8):1115.
3. Dalawai SP, Saad Aly MA, Latthe SS, Xing R, Sutar RS, Nagappan S, et al. Recent Advances in durability of superhydrophobic self-cleaning technology: A critical review. *Prog Org Coatings* [Internet]. 2020;138(Sep 2019):105381. Available from: https://doi.org/10.1016/j.porgcoat.2019.105381
4. Mozumder MS, Mourad AHI, Pervez H, Surkatti R. Recent developments in multifunctional coatings for solar panel applications: A review. *Sol Energy Mater Sol Cells* [Internet]. 2019;189(June 2018):75–102. Available from: https://doi.org/10.1016/j.solmat.2018.09.015
5. Rajbahadur YNK, Kumar A, Negi S, Kar S. Evaluation of hydrophobic/hydrophilic and antireflective coatings for photovoltaic panels. *J Coatings Technol Res* [Internet]. 2024. Available from: https://doi.org/10.1007/s11998-024-00929-0
6. Collins CM, Safiuddin M. Lotus-leaf-inspired biomimetic coatings: Different types, key properties, and applications in infrastructures. *Infrastructures*. 2022 Mar 23;7(4):46.
7. Catching water with imitation beetle bumps | Research | Chemistry World [Internet]. [cited 2024 Jul 7]. Available from: https://www.chemistryworld.com/news/catching-water-with-imitation-beetle-bumps/8091.article

8. Darmanin T, Guittard F. Superhydrophobic and superoleophobic properties in nature. *Mater Today*. 2015 Jun;18(5):273–85.
9. Zheng Y, Gao X, Jiang L. Directional adhesion of superhydrophobic butterfly wings. *Soft Matter*. 2007;3(2):178–82.
10. Niu S, Li B, Mu Z, Yang M, Zhang J, Han Z, et al. Excellent structure-based multifunction of morpho butterfly wings: A review. *J Bionic Eng*. 2015 Jun 1;12(2):170–89.
11. Neinhuis C. Characterization and distribution of water-repellent, self-cleaning plant surfaces. *Ann Bot*. 1997 Jun;79(6):667–77.
12. Nakajima A, Hashimoto K, Watanabe T. Recent studies on super-hydrophobic films. *In: Molecular Materials and Functional Polymers*. Vienna: Springer Vienna; 2001. p. 31–41.
13. Sun T, Feng L, Gao X, Jiang L. Bioinspired surfaces with special wettability. *Acc Chem Res*. 2005 Aug 1;38(8):644–52.
14. Parkin IP, Palgrave RG. Self-cleaning coatings. *J Mater Chem*. 2005;15(17):1689.
15. Rajbahadur YNK, Kumar A, Kumar MM, Negi S, Kar S. A critical review on anti-soiling and anti-reflective coatings for self-cleaning surface on solar photovoltaic panels. *In: International Conference on Recent Advancements in Mechanical Engineering*. Singapore: Springer Nature Singapore; 2025. p. 153–69.
16. Ahmad D, van den Boogaert I, Miller J, Presswell R, Jouhara H. Hydrophilic and hydrophobic materials and their applications. *Energy Sources, Part A Recover Util Environ Eff*. 2018 Nov 17;40(22):2686–725.
17. Kim D, Kim W, Park S, Kim S, Gwon Y, Kim J. Leaf-inspired micro- and nanoengineered surfaces for controlled hydrophilic and hydrophobic properties. *Macromol Res*. 2020 Jan 29;28(1):57–61.
18. Liu X, Liang Y, Zhou F, Liu W. Extreme wettability and tunable adhesion: Biomimicking beyond nature? *Soft Matter*. 2012;8(7):2070–86.
19. Zhu H, Guo Z, Liu W. Biomimetic water-collecting materials inspired by nature. *Chem Commun*. 2016;52(20):3863–79.
20. Adak D, Bhattacharyya R, Barshilia HC. A state-of-the-art review on the multifunctional self-cleaning nanostructured coatings for PV panels, CSP mirrors and related solar devices. *Renew Sustain Energy Rev* [Internet]. 2022;159(February):112145. Available from: https://doi.org/10.1016/j.rser.2022.112145
21. Dahlioui D, Laarabi B, Barhdadi A. Review on dew water effect on soiling of solar panels: Towards its enhancement or mitigation. *Sustain Energy Technol Assessments* [Internet]. 2022;49(Aug 2021):101774. Available from: https://doi.org/10.1016/j.seta.2021.101774
22. Deb D, Brahmbhatt NL. Review of yield increase of solar panels through soiling prevention, and a proposed water-free automated cleaning solution. *Renew Sustain Energy Rev* [Internet]. 2018;82(Apr 2017):3306–13. Available from: https://doi.org/10.1016/j.rser.2017.10.014
23. Yao L, He J. Recent progress in antireflection and self-cleaning technology – From surface engineering to functional surfaces. *Prog Mater Sci* [Internet]. 2014;61:94–143. Available from: http://dx.doi.org/10.1016/j.pmatsci.2013.12.003
24. Hosseini MS, Ebratkhahan M, Shayegan Z, Niaei A, Salari D, Rostami A, et al. Investigation of the effective operational parameters of self-cleaning glass surface coating to improve methylene blue removal efficiency; application in solar cells. *Sol Energy* [Internet]. 2020;207(July):398–408. Available from: https://doi.org/10.1016/j.solener.2020.06.109
25. Syafiq A, Pandey AK, Adzman NN, Rahim NA. Advances in approaches and methods for self-cleaning of solar photovoltaic panels. *Sol Energy*. 2018;162(May 2017):597–619.
26. Jain R, Pitchumani R. Fractal model for wettability of rough surfaces. *Langmuir*. 2017 Jul 18;33(28):7181–90.

27. Hirasakl GJ. Wettability: Fundamentals and surface forces. *SPE Form Eval.* 1991 Jun 1;6(02):217–26.
28. Stammitti-Scarpone A, Acosta EJ. Solid-liquid-liquid wettability and its prediction with surface free energy models. *Adv Colloid Interface Sci.* 2019 Feb;264:28–46.
29. Anderson W. Wettability literature survey- Part 2: Wettability measurement. *J Pet Technol.* 1986 Nov 1;38(11):1246–62.
30. Lashgari HR, Xu Y, Sepehrnoori K. Modeling dynamic wettability alteration effect based on contact angle. *In: All Days.* SPE; 2016.
31. Wenzel RN. Resistance of solid surfaces to wetting by water. *Ind Eng Chem.* 1936 Aug 1;28(8):988–94.
32. Yen TH, Soong CY. Hybrid Cassie-Wenzel model for droplets on surfaces with nanoscale roughness. *Phys Rev E.* 2016 Feb 25;93(2):022805.
33. Bell MS, Borhan A. A volume-corrected Wenzel model. *ACS Omega.* 2020 Apr 21;5(15):8875–84.
34. Hao JH, Wang ZJ. Modeling cassie–Baxter state on superhydrophobic surfaces. *J Dispers Sci Technol.* 2016 Aug 2;37(8):1208–13.
35. Chang Z, Lu Y. Fabrication of superhydrophobic surfaces with Cassie-Baxter state. *J Dispers Sci Technol.* 2022 Jun 15;43(8):1099–111.
36. Liu TL, Chen Z, Kim CJ. A dynamic Cassie–Baxter model. *Soft Matter.* 2015;11(8):1589–96.
37. Kumar A, Rajbahadur YNK, Kumar MM, Negi S, Kar S. A critical review on the fabrication of superhydrophobic membrane for oil–water separation. *In: International Conference on Recent Advancements in Mechanical Engineering.* Singapore: Springer Nature Singapore; 2025. p. 125–36.
38. Raut HK, Ganesh VA, Nair AS, Ramakrishna S. Anti-reflective coatings: A critical, in-depth review. *Energy Environ Sci.* 2011;4(10):3779–804.
39. Du X, Xing Y, Zhou M, Li X, Huang H, Meng XM, et al. Broadband antireflective superhydrophilic antifogging nano-coatings based on three-layer system. *Microporous Mesoporous Mater* [Internet]. 2018;255:84–93. Available from: https://doi.org/10.1016/j.micromeso.2017.07.017
40. Wu X, Wyman I, Zhang G, Lin J, Liu Z, Wang Y, et al. Preparation of superamphiphobic polymer-based coatings via spray- and dip-coating strategies. *Prog Org Coatings.* 2016;90:463–71.
41. Sriram S, Singh RK, Kumar A. Silica and Silane based polymer composite coating on glass slide by dip-Coating Method. *Surfaces and Interfaces* [Internet]. 2020;19(Feb):100472. Available from: https://doi.org/10.1016/j.surfin.2020.100472
42. Wang SD, Wang JY. Anti-reflective and superhydrophobic films prepared from a sol at different withdrawal speeds. *Appl Surf Sci.* 2019 May;476:1035–48.
43. Topçu Kaya AS, Cengiz U. Fabrication and application of superhydrophilic antifog surface by sol-gel method. *Prog Org Coatings* [Internet]. 2019;126(Oct 2018):75–82. Available from: https://doi.org/10.1016/j.porgcoat.2018.10.021
44. Biswas PK, Sujatha Devi P, Chakraborty PK, Chatterjee A, Ganguli D, Kamath MP, et al. Porous anti-reflective silica coatings with a high spectral coverage by sol-gel spin coating technique. *J Mater Sci Lett.* 2003;22(3):181–3.
45. Nisticò R, Scalarone D, Magnacca G. Sol-gel chemistry, templating and spin-coating deposition: A combined approach to control in a simple way the porosity of inorganic thin films/coatings. *Microporous Mesoporous Mater.* 2017 Aug 1;248:18–29.
46. Nomeir B, Lakhouil S, Boukheir S, Ali MA, Naamane S. Recent progress on transparent and self-cleaning surfaces by superhydrophobic coatings deposition to optimize the cleaning process of solar panels. *Sol Energy Mater Sol Cells* [Internet]. 2023 Aug 1 [cited 2023 May 16];257:112347. Available from: https://linkinghub.elsevier.com/retrieve/pii/S092702482300168X

47. Sutha S, Suresh S, Raj B, Ravi KR. Transparent alumina based superhydrophobic self–cleaning coatings for solar cell cover glass applications. *Sol Energy Mater Sol Cells* [Internet]. 2017;165(Jan):128–37. Available from: http://dx.doi.org/10.1016/j.solmat.2017.02.027
48. Cui Y, Yu H, Abbas Z, Wang Z, Wang L, Wang D. PZT composite film preparation and characterization using a method of sol-gel and electrohydrodynamic jet printing. *Micromachines*. 2023 Apr 24;14(5):918.
49. Sumanth Kumar D, Jai Kumar B, Mahesh HM. Quantum Nanostructures (QDs): An overview. *In: Synthesis of Inorganic Nanomaterials*. Elsevier; 2018. p. 59–88.
50. Jung DS, Park S Bin, Kang YC. Design of particles by spray pyrolysis and recent progress in its application. *Korean J Chem Eng*. 2010 Nov 21;27(6):1621–45.
51. Mooney JB, Radding SB. Spray pyrolysis processing. *Annu Rev Mater Sci*. 1982 Aug;12(1):81–101.
52. Tarwal NL, Patil PS. Superhydrophobic and transparent ZnO thin films synthesized by spray pyrolysis technique. *Appl Surf Sci*. 2010 Oct;256(24):7451–6.
53. Dahamni MA, Ghamnia M, Naceri SE, Fauquet C, Tonneau D, Pireaux JJ, et al. Spray pyrolysis synthesis of pure and Mg-doped manganese oxide thin films. *Coatings*. 2021 May 19;11(5):598.
54. Frey H. Chemical vapor deposition (CVD). *Handb Thin-Film Technol*. 2015;225–52.
55. Jeong YM, Lee JK, Jun HW, Kim G rak, Choe Y. Preparation of super-hydrophilic amorphous titanium dioxide thin film via PECVD process and its application to dehumidifying heat exchangers. *J Ind Eng Chem*. 2009 Mar;15(2):202–6.
56. Jatoi AS, Akhter F, Mubarak NM, Mazari SA, Ahmed S, Sabzoi N, et al. Conventional techniques for nanomaterials preparation. *Sustain Nanotechnol Environ Remediat*. 2022 Jan 1;91–110.
57. Huang ZS, Shen C, Fan L, Ye X, Shi X, Li H, et al. Experimental investigation of the anti-soiling performances of different wettability of transparent coatings: Superhydrophilic, hydrophilic, hydrophobic and superhydrophobic coatings. *Sol Energy Mater Sol Cells* [Internet]. 2021;225(March):111053. Available from: https://doi.org/10.1016/j.solmat.2021.111053
58. Sabzi M, Mousavi Anijdan S, Shamsodin M, Farzam M, Hojjati-Najafabadi A, Feng P, et al. A review on sustainable manufacturing of ceramic-based thin films by chemical vapor deposition (CVD): Reactions kinetics and the deposition mechanisms. *Coatings*. 2023 Jan 14;13(1):188.
59. Lin SS, Huang JL, Šjgalik P. The properties of Ti-doped ZnO films deposited by simultaneous RF and DC magnetron sputtering. *Surf Coatings Technol*. 2005 Feb 21;191(2–3):286–92.
60. Pan TJ, Dai YJ, Jiang J, Xiang JH, Yang QQ, Li YS. Anti-corrosion performance of the conductive bilayer CrC/CrN coated 304SS bipolar plate in acidic environment. *Corros Sci*. 2022 Sep 1;206:110495.
61. Seong SG, Kim EJ, Kim YS, Lee KE, Hahn SH. Influence of deposition atmosphere on photocatalytic activity of TiO2/SiOx double-layers prepared by RF magnetron sputtering. *Appl Surf Sci*. 2009 Oct 15;256(1):1–5.
62. Ellmer K, Cebulla R, Wendt R. Transparent and conducting ZnO(:Al) films deposited by simultaneous RF- and DC-excitation of a magnetron. *Thin Solid Films*. 1998 Apr 1;317(1–2):413–6.
63. Liu YY, Qian LQ, Guo C, Jia X, Wang JW, Tang WH. Natural superhydrophilic TiO2/SiO2 composite thin films deposited by radio frequency magnetron sputtering. *J Alloys Compd*. 2009 Jun;479(1–2):532–5.
64. Maurya D, Sardarinejad A, Alameh K. Recent Developments in R.F. Magnetron sputtered thin films for pH sensing applications – An overview. *Coatings*. 2014 Dec 1;4(4):756–71.

65. Shahzadi P, Gilani SR, Rana BB, Ghaffar A, Munir A. Transparent, self-cleaning, scratch resistance and environment friendly coatings for glass substrate and their potential applications in outdoor and automobile industry. *Sci Rep.* 2021 Oct 20;11(1):20743.
66. Chandru K, Elansezhian R. A study on impact of hydrophobic effect on Al2O3 coated glass by sol–gel dip coating method for automobile windshield application. *In: International Conference on Transportation Infrastructure Projects: Conception to Execution.* 2023; p. 77–85.
67. Bai Y, Zhang H, Shao Y, Zhang H, Zhu J. Recent progresses of superhydrophobic coatings in different application fields: An overview. *Coatings.* 2021 Jan 20;11(2):116.
68. Gozutok Z, Kinj O, Torun I, Ozdemir AT, Onses MS. One-step deposition of hydrophobic coatings on paper for printed-electronics applications. *Cellulose.* 2019 Mar 25;26(5):3503–12.
69. A. KS, P. D, G. D, J. N, G.S. H, S. AS, et al. Super-hydrophobicity: Mechanism, fabrication and its application in medical implants to prevent biomaterial associated infections. *J Ind Eng Chem.* 2020 Dec;92:1–17.
70. Xie J, Hu J, Fang L, Liao X, Du R, Wu F, et al. Facile fabrication and biological properties of super-hydrophobic coating on magnesium alloy used as potential implant materials. *Surf Coatings Technol.* 2020 Feb;384:125223.
71. Sethi SK, Manik G. Recent progress in super hydrophobic/hydrophilic self-cleaning surfaces for various industrial applications: A review. *Polym Plast Technol Eng.* 2018 Dec 12;57(18):1932–52.
72. Buczkowska K. Hydrophobic protection for building materials. *In: Superhydrophobic Coating – Recent Advances in Theory and Applications.* IntechOpen; 2023.
73. Wang Y, Huang Y. Modification of hydrophobicity of concrete by nanocomposite particles. *Integr Ferroelectr.* 2021 Mar 24;215(1):233–48.
74. Faraldos M, Kropp R, Anderson MA, Sobolev K. Photocatalytic hydrophobic concrete coatings to combat air pollution. *Catal Today.* 2016 Jan;259:228–36.
75. Bahners T, Textor T, Opwis K, Schollmeyer E. Recent approaches to highly hydrophobic textile surfaces. *J Adhes Sci Technol.* 2008 Jan;22(3–4):285–309.
76. Mahltig B. Hydrophobic sol–gel-based coating agent for textiles: Improvement by solvothermal treatment. *J Text Inst.* 2011 May;102(5):455–9.
77. Ruzi M, Celik N, Onses MS. Superhydrophobic coatings for food packaging applications: A review. *Food Packag Shelf Life.* 2022 Jun;32:100823.
78. Asim N, Badiei M, Mohammad M. Recent advances in cellulose-based hydrophobic food packaging. *Emergent Mater.* 2022 Jun 4;5(3):703–18.
79. Piscitelli F, Tescione F, Mazzola L, Bruno G, Lavorgna M. On a simplified method to produce hydrophobic coatings for aeronautical applications. *Appl Surf Sci.* 2019 Apr;472:71–81.
80. Huang X, Tepylo N, Pommier-Budinger V, Budinger M, Bonaccurso E, Villedieu P, et al. A survey of icephobic coatings and their potential use in a hybrid coating/active ice protection system for aerospace applications. *Prog Aerosp Sci.* 2019 Feb;105:74–97.
81. Uddin MN, Desai FJ, Rahman MM, Asmatulu R. A highly efficient fog harvester of electrospun permanent superhydrophobic–hydrophilic polymer nanocomposite fiber mats. *Nanoscale Adv.* 2020;2(10):4627–38.
82. Seo D, Lee J, Lee C, Nam Y. The effects of surface wettability on the fog and dew moisture harvesting performance on tubular surfaces. *Sci Rep.* 2016 Apr 11;6(1):24276.
83. Baig U, Shaukat MM, Shuja SZ, Asif M, Khan NA. Development and life cycle assessment (LCA) of super-oleophobic (under water) and super-hydrophilic (in-air) mesh membrane for oily water treatment. *Sci Rep.* 2024 Jul 3;14(1):15268.
84. Knapczyk-Korczak J, Szewczyk PK, Ura DP, Berent K, Stachewicz U. Hydrophilic nanofibers in fog collectors for increased water harvesting efficiency. *RSC Adv.* 2020;10(38):22335–42.

85. Wu P, Xue Z, Yu T, Penkov O. Transparent self-cleaning coatings: A review. *Coatings*. 2023 Jul 19;13(7):1270.
86. Wang B, Zhou X, Guo Z, Liu W. Recent advances in atmosphere water harvesting: Design principle, materials, devices, and applications. *Nano Today*. 2021 Oct;40:101283.
87. Ahmad M, Nighojkar A, Plappally A. A review of the methods of harvesting atmospheric moisture. *Environmental Science Pollution Research*. 2024;31(7):10395–416.
88. Thakur AK, Sathyamurthy R, Velraj R, Saidur R, Lynch I, Chaturvedi M, et al. Synergetic effect of absorber and condenser nano-coating on evaporation and thermal performance of solar distillation unit for clean water production. *Sol Energy Mater Sol Cells*. 2022 Jun;240:111698.
89. Chen J, Yin JL, Li B, Ye Z, Liu D, Ding D, et al. Janus evaporators with self-recovering hydrophobicity for salt-rejecting interfacial solar desalination. *ACS Nano*. 2020 Dec 22;14(12):17419–27.
90. Liu C, Lin Z, Qiao C, Zhao Z, Wang C, Sun X, et al. Hemocompatibility assay of a micro-catheter using hydrophilic coating biomaterials. *Biomed Mater Eng*. 2019 Jan 24;30(1):1–9.
91. Wyman P. Hydrophilic coatings for biomedical applications in and ex vivo. *In: Coatings for Biomedical Applications*. Elsevier; 2012. p. 3–42.
92. Chen Y, Yan H, Pan Y. Wetting and evaporative performance analysis of wet channels in indirect evaporative cooler with hydrophilic nano-coating. *Appl Therm Eng*. 2023 Jul;229:120622.
93. Qi R, Hu Y, Wang Y, Lu L. A new approach to enhance the heat and mass transfer of liquid desiccant dehumidification with a titanium dioxide superhydrophilic self-cleaning coating. *J Clean Prod*. 2016 Jan;112:3555–61.
94. Vladisavljević GT, Khalid N, Neves MA, Kuroiwa T, Nakajima M, Uemura K, et al. Industrial lab-on-a-chip: Design, applications and scale-up for drug discovery and delivery. *Adv Drug Deliv Rev*. 2013 Nov;65(11–12):1626–63.
95. Sonntag F, Grünzner S, Schmieder F, Busek M, Klotzbach U, Franke V. Multilayer based lab-on-a-chip-systems for substance testing. *In: Laser-based Micro-and Nanoprocessing IX*. 2015; p. 48–57.
96. Xu Y, Takai M, Ishihara K. Functional coatings for lab-on-a-chip systems based on phospholipid polymers. *In: Handbook of Modern Coating Technologies*. Elsevier; 2021. p. 555–95.

7 Spray Pyrolysis Deposition of Thin Films
A Comprehensive Review

Aswathy M. Narayanan, Avala Ramesh, Krishnadas Narayanan Nampoothiri, and Sukanti Behera

7.1 INTRODUCTION

Thin films are layers of material supported on a substrate whose thickness ranges from a few nanometres to several micrometres [1]. Mainly, the deposition technique [2] can be classified into two: physical and chemical processes [3]. The physical deposition techniques, such as thermal evaporation and sputtering, require complicated and expensive equipment and are characterized by large material wastage during the deposition process [4]. It is also at a disadvantage when considering large-area deposition since most of these techniques are carried out under vacuum. Whereas chemical deposition methods such as dip coating and spray pyrolysis deposition are largely simple and cost-effective and can be used to deposit thin films over a large area [5]. The selection of appropriate methods depends on the thickness, substrate, and the specific properties required.

Among the various chemical deposition techniques, spray pyrolysis deposition (SPD) is a widely used method [6] because of its simplicity as well as its ability to produce uniform thin films over a large area. A variety of conducting, semiconducting, and insulating materials like metal chalcogenides [7], metal oxides [8], alloys [9, 10], and nanocomposites [11] have been deposited in thin and thick film forms utilizing this versatile technique. Doping is extremely easy with SPD because the dopant can be simply incorporated into the spray solution in any concentration [12, 13]. The method being an open atmosphere process can function without the vacuum conditions, and it does not require high-quality substrates to enable industrial applications [14]. Dense or porous thin films of various materials can be deposited by controlling the deposition conditions. Controlling the morphology of the deposited films is also possible by precise control of the deposition conditions [15]. Multi-layered materials or compositionally graded thin films can also be prepared using the severe plastic deformation (SPD) method by a sequential deposition procedure [16].

This chapter provides a comprehensive review of thin film materials deposited using spray pyrolysis. A discussion on the instrument and its components is followed by the types of spray pyrolysis techniques such as ultrasonic spray pyrolysis (USP), electrostatic spray pyrolysis (ESP), flame spray pyrolysis (FSP), and pressurized

 DOI: 10.1201/9781003750659-7

spray pyrolysis (PSP). Various factors that affect the thin film deposition processes are discussed thereafter. A brief overview of different materials deposited by SPD and their applications is provided. Finally, the chapter concludes with a summary and challenges in this field.

7.2 INSTRUMENTATION

During SPD, thin films are created by spraying precursor solutions of certain materials onto a heated substrate [16]. A precursor solution is atomized using various techniques to obtain a mist which is transported to the hot substrate. During the flight of the mist towards the hot substrate, solvent evaporation and the subsequent shrinkage of droplets occur [17]. The mist hits the hot substrate to form a disk-like structure, and the subsequent solvent evaporation and/or pyrolysis causes the formation of films of the desired material [18]. The factors influencing the disk's size and shape are the substrate temperature and the droplet's volume and momentum [15]. Figure 7.1 represents the schematic diagram of the deposition process. The deposition set-up mainly consists of (i) nozzle for spray deposition, (ii) container containing the precursor solution, (iii) substrate, (iv) heater attached to the substrate, (v) temperature controller device to control the temperature, (vi) air compressor, (vii) gas-regulating valve, and (viii) electronic controller components. Various flow meters are used to quantify the flow rates of compressed air and the precursor solutions. Also, the substrate or the nozzle is moved in a controlled manner to achieve uniform thin films. Sometimes, both the nozzle and substrate are moved to attain uniform deposition films. The movement of the spray assembly is controlled using a moving table which is connected to stepping motors, and the spray is restored over the substrate for uniform film deposition [14, 15, 18].

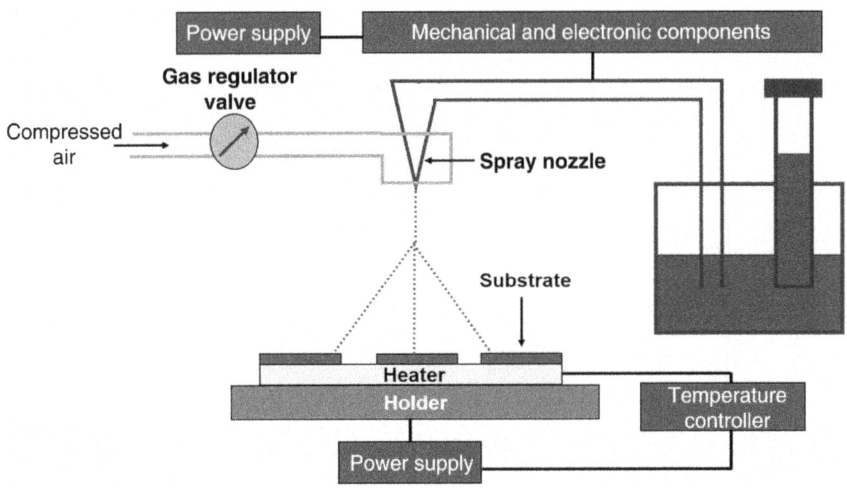

FIGURE 7.1 Schematic diagram depicting the deposition process.

7.3 CLASSIFICATION

Depending upon the method to create fine droplets or aerosols of the precursor solution, SPD can be classified into ultrasonic spray pyrolysis (USP) [19–21], electrostatic spray pyrolysis (ESP) [11, 22], flame spray pyrolysis (FSP) [11, 15, 23, 24], and pressurized spray pyrolysis (PSP) [13, 25]. USP utilizes vibrational acceleration, ESP uses electrostatic acceleration, FSP operates at high temperature, and PSP employs pneumatic acceleration for atomization and acceleration of the droplets.

The ultrasonic spray pyrolysis method (USP) is one of the spray pyrolysis methods which uses an economically viable and malleable technique of aerosol process for nanoparticle synthesis and thin film deposition. By vibrating at high frequencies, ultrasonic waves are produced [11, 14], and when the precursor solution is subjected to such waves, aerosol droplets are formed due to atomization of the solution. These fine droplets will be guided to a hot chamber using a carrier gas where the droplets will undergo several processes such as drying, contraction, precipitation, thermalization, and sintering to form spherical particles [6] (see Figure 7.2).

In the ESP technique, electrohydrodynamic (EHD) force is utilized to generate the spray of the precursor solution. DC voltage, in the range of kVs, is applied between an electrically conductive substrate and a metal capillary nozzle. The precursor solution is linked to the nozzle, and a peristaltic pump is utilized to control the flow rate of the solution (see Figure 7.3). Due to the electrostatic force generated during the process, the spray moves towards the heated substrate. On the heated substrate, pyrolysis of the precursors occurs, and a thin layer is deposited on the substrate surface. Due to its cheap and simple process, this process is used for synthesizing powders of sizes in the order of sub-microns.

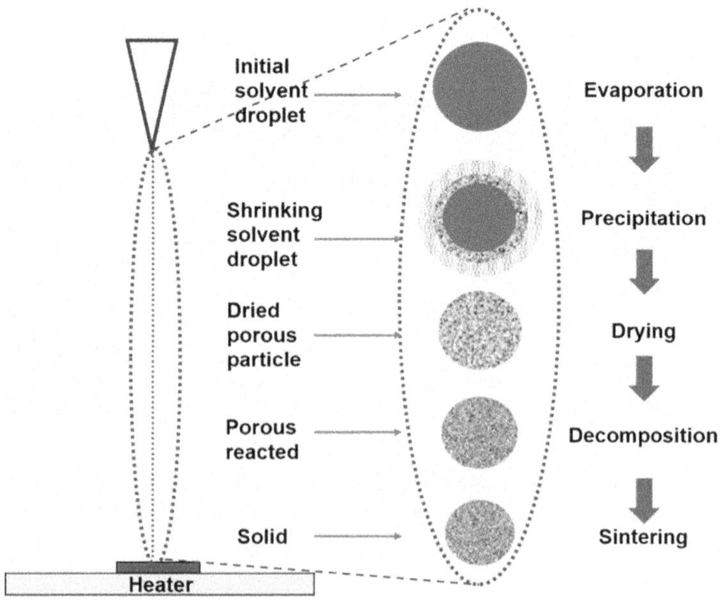

FIGURE 7.2 Schematic diagram showcasing various processes.

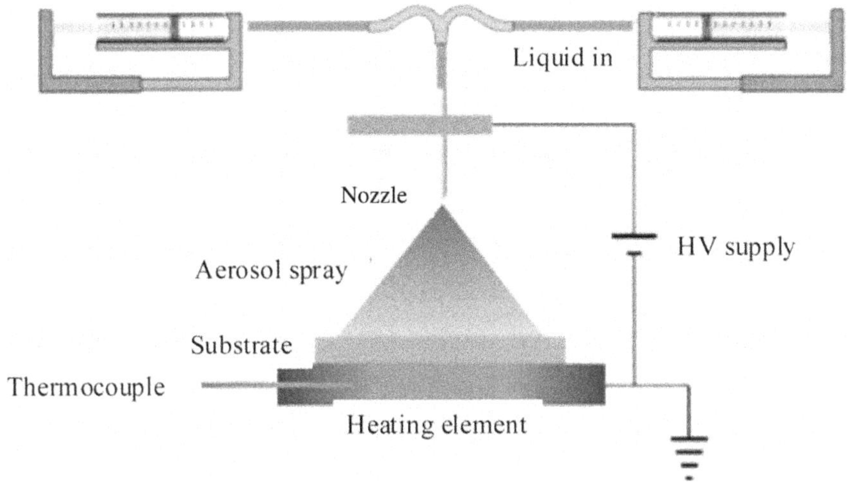

FIGURE 7.3 Schematic diagram of ESP technology [26].

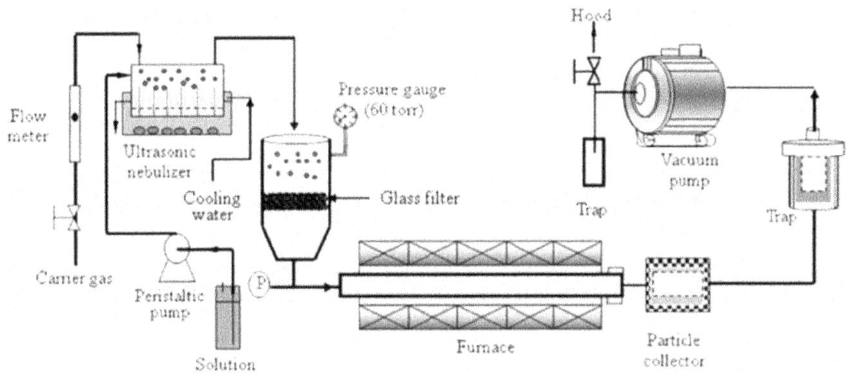

FIGURE 7.4 Schematic diagram portraying filter expansion aerosol generator process [32].

PSP technique is another type of spray pyrolysis where micrometre-sized droplets are produced under the application of pressure. Filter expansion aerosol generator is a kind of PSP, and it consists of various components such as a reactor, particle collector, atomizer, and pressure controller as shown in Figure 7.4 and Figure 7.5. Through the nozzle, the liquid is sprayed with the aid of a carrier gas onto the glass filter surface which results in the condensation of the liquid to a thin coating. This method is used to form nanoparticles directly from metals, metal oxides, and composite materials. The precursor solutions of metal or metal oxides are prepared by dissolving in a solvent and spraying the solution into a high-temperature reactor. During the process, inside the reactor thermal decomposition occurs through high pressure to form nanoparticles. Due to the difference in pressure in the reactor, the liquid film expands to form jets and subsequent condensation to droplets. Various

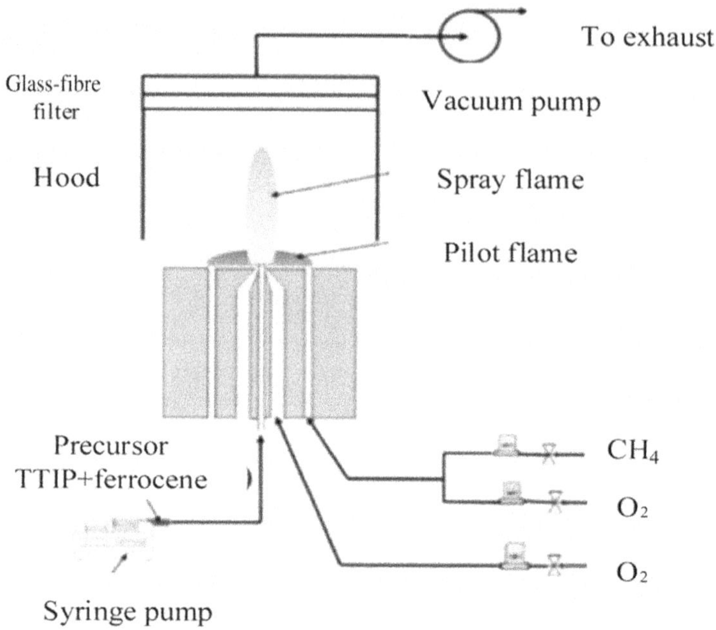

FIGURE 7.5 Schematic diagram depicting FSP technology [30].

factors such as precursor concentration, spray flow rate, reactor conditions such as temperature and pressure, and the occurrence of additives affect the size and shape of the nanoparticles.

The FSP process comprises various components which are used to perform certain functions such as a spray process and for delivery and collection of the particles [27–29]. The spray system consists of a regulated spray nozzle. When the precursor solution is passed through a capillary tube, the dispersion gas is allowed to flow through the annular gap. The spray nozzle is fed with oxygen dispersion gas at a certain pressure drop for flame production. An annular slit surrounding the nozzle provides the supporting premixed pilot flame for methane (CH_4). When the precursor solution is subjected to the flame, atomization into tiny droplets occurs. Along with thin film synthesis, this process is also used for nanoparticle and metal oxide synthesis [23, 30, 31].

7.4 EFFECT OF PROCESS PARAMETERS

As already mentioned, the aerosol particles created during atomization are transferred to the hot substrate to undergo thermal decomposition during spray pyrolysis deposition. These droplets hit the substrate to form a disk-shaped structure, the shape and size of which depends on the deposition conditions like a) substrate temperature, b) solution flow rate and rate of deposition, c) precursor solution, d) additives, e) droplet size, and f) the distance between nozzle and substrate [11, 14, 15]. These parameters greatly influence the morphology, surface roughness, thickness, and physical and chemical properties of the thin film [17, 33–35]. Even though it is difficult to generalize the effect of deposition parameters on the quality of the film, an attempt is made to discuss the effect of the aforementioned parameters on the overall quality of the deposited thin films in this section.

7.4.1 Substrate Temperature

Multiple steps such as atomization of the precursor solution and transport, solvent evaporation, droplet impact with the substrate, thermal decomposition, and sintering occur during SPD [17]. Apart from the atomization and transport step, all the other steps are affected by the substrate temperature. Thus, the thin films produced with varying substrate temperatures are expected to possess different characteristics. A low substrate temperature may yield an amorphous or powdery film, whereas a high substrate temperature may give rise to well-crystalline, porous, and rough films. Thus, optimization of substrate temperature is necessary to obtain the necessary chemical and physical characteristics for a given material and thickness.

Alver et al. [20] deposited [002] oriented ZnO microrods of varying diameters by changing the deposition temperature from 400 °C to 550 °C. The optical transmittance and bandgap energies increased with an increase in deposition temperature. Zaier et al. [36] studied the effect of substrate temperature on the structural and optoelectrical properties of ZnO thin films deposited by SPD on a glass substrate. They found that the transmittance as well as the conductivity of the films increase with an increase in deposition temperature from 250 °C to 450 °C. However, better crystalline films were obtained at a moderate substrate temperature of 350 °C. This may be due to an increase in crystal defects at high deposition temperatures.

Firat and coworkers [21] investigated the effect of substrate temperature on the properties of CuS thin films deposited by USP. There was a clear change in the crystal structure of the films from hexagonal to cubic on increasing the deposition temperature from 240 °C to 320 °C. The Cu/S ratio was also found to increase with an increase in substrate temperature. Both crystallite size and surface roughness are reduced with increasing deposition temperature. Bian et al. [19] reported a change in the film growth mechanism by increasing the substrate temperature from 580 °C to 680 °C on MgO thin films deposited by USP. Lower temperatures (580 °C) resulted in an island growth mechanism, whereas higher temperatures (680 °C) resulted in a layer growth mechanism as observed from the varied morphology.

7.4.2 Solution Flow Rate and Rate of Deposition

The solution flow rate decides how much time is available for a drop to land on the surface and react before another drop strikes the surface. This parameter influences the nucleation and subsequent growth of the films and thus is critical in deciding the morphology of the films. A slow solution flow rate and low surface temperature may result in a film with an aqueous (or solvent) residing film. This may be because the conditions create an atmosphere saturated with water (or solvent) vapour due to less time available for solvent evaporation.

Zeggar et al. [37] studied the effect of flow rate on the properties of spray pyrolysis-coated CuO thin film. They found that above a certain critical flow rate, the film thickness varied linearly with flow rate. The texture of the film varied with flow rate; an inversion in preferred orientation was observed with variation in flow rate. A low flow rate (< 25 ml/h) gave rise to (002) orientation while a flow rate larger than 25 ml/h resulted in (111) preferred orientation. Crystallite size, bandgap, and optical transmittance decreased with an increase in flow rate.

7.4.3 Precursor Solution

The type of solvent employed for dissolving the metal precursors changes the boiling point, solubility, and spreading behaviour of the droplets [30]. This affects the deposition temperature, maximum salt concentration, and optimal flow rate, eventually altering the deposition rate. Mostly, spray pyrolysis processes use aqueous solvents and a variety of metal salts like chlorides, acetates, formates, and nitrates. Organic solvents like ethanol, methanol, or acetone with lower boiling points than water have also been reported as solvents for SPD. Alcoholic solvents are preferred due to their low surface tension and viscosity, which facilitates the formation of small droplets. The low boiling point of organic solvents enables them to be efficiently removed from the deposition chamber in the vapour phase [38]. Most of the organic solvents get evaporated during the transport and before striking the surface, leaving the vapours of the precursors to react on the surface.

C. Chen et al. [39] discussed the effect of solvent on the morphology of $LiCoO_2$ thin films deposited using electrostatic spray pyrolysis. They have employed pure ethanol and a mixture of ethanol and butyl carbitol in various compositions to understand the effect of solvents. The boiling points of ethanol and butyl carbitol are 78 °C and 230 °C respectively. They have found that the porosity of the films altered by changing the solvent composition at a similar deposition temperature. For instance, at a deposition temperature of 250 °C the films obtained were pretty dense when a mixture of 50 volume% ethanol and 50 volume% butyl carbitol was used. The films deposited with pure ethanol as solvent were porous. High boiling point solvents result in a slower precipitation step compared to low boiling solvents because the droplets are still wet when they reach the hot substrate, creating a dense morphology.

The concentration of the spray solution is an important variable affecting the morphology of the films. Low solution concentrations along with slow deposition rates have been found to give columnar grains and smooth films, whereas higher solution concentration gives rise to rough films. The solution concentration also affects the thickness of the films affecting their electrical and optical properties.

B. A. Victor [40] reported the effect of precursor solution concentration on the optical properties of ZnO thin films. The precursor solution concentration was varied from 0.01 M to 0.05 M. The thickness of the films increased with an increase in precursor solution concentration which in turn reduced the optical transparency of the films. A similar effect was observed by Owoeye et al., the crystallite size increased and the optical transmittance reduced on increasing the spray solution concentration from 0.1 M to 0.3 M. They have also observed an enhanced corrosion resistance of the ZnO-coated stainless steel substrates on increasing the precursor concentration.

Apart from the concentration of the precursor, the type of the precursor also affects the film properties. The metal salt precursor should dissolve in water or in any common solvent and should decompose at the deposition temperature to give the corresponding oxide or react with other precursors to give the required thin film [38]. Inorganic metal salts like nitrates and chlorides are commonly used as precursors [41, 42]. Organometallic compounds and metal-organic salts have higher volatility compared to inorganic metal salts [43, 44]. This changes the morphology of the deposited films because of the gaseous to solid state conversion on transfer to the substrate. Colloids like sols and emulsions alter the film porosity [45, 46].

Attouche et al. [47] have reported the effect of the metal salt on the properties of the TiO_2 thin films prepared by SPD. They used two metal precursors, namely titanium tetra isopropoxide (TTIP) and titanium chloride, for depositing thin films under identical conditions. In both cases, the thickness increased progressively with increasing the concentration of the spray solution. Smoother and crystalline films were obtained with the TTIP precursor, whereas the chloride precursor gave rise to comparatively rougher films with globular grains.

7.4.4 Additives

To modify the morphology of the deposited film, the literature reports the addition of additives, especially polymeric additives into the starting precursor solution. Surfactants such as sodium dodecyl sulphate (SDS) and polyvinylpyrrolidone (PVP) are used in USP to modify the surface tension and viscosity of the deposited film. Salts like NaCl and sucrose are added to the precursor solution to prevent agglomeration. To enhance the electrical properties such as electrical conductivity, caron nanotubes (CNTs) are being added to the solution.

Sanchez et al. [48] achieved non-cracked nanostructured cerium dioxide thin films by using acetic acid as an additive. The researchers have used the USP method to deposit the prepared nanostructured cerium dioxide thin films onto a single crystalline silicon substrate. Stopic et al. [49] have prepared non-agglomerated spherical Ni particles through the USP method using Cu, Pd, and Ni as additives. Characterization of the particles was done using thermogravimetric analysis and differential thermal analysis (DTA) to observe the presence of the additives.

7.4.5 Droplet Size

As mentioned in the previous section, SPD method is utilized to deposit thin films by spraying the precursor solution over the heated substrate. Upon spraying, the generated droplets will come in contact with the substrate thereby transforming into a disk-like structure, which is followed by thermal synthesis [14]. Various factors such as temperature of the substrate, droplet volume, and droplet momentum influence the size and shape of the disk. This results in overlapping disks of the precursor solution, which gets oxidized on the heated substrate [11]. Thus, the morphological properties of the deposited film are recognized by droplet size distribution. Larger precursor drops create particles with holes due to high solvent evaporation rates [15]. Tsai et al. [50] analysed the influence of various parameters such as drop size and precursor concentration on the deposited film by USP. The nebulizers were operated at 1.65 MHz and 23.5 W, and thus droplets with diameters in the range 6–9 mm were generated. This resulted in the production of monodisperse spherical particles of diameter 90 nm. Pradita et al. [51] reported that hollow, spherical, and irregular MgO particles can be produced from MgN at 700 °C, 800 °C, and 900 °C using spray pyrolysis. They inferred that with an increase in SP temperatures, the particle size decreases, and the achieved minimum particle size was in the range of 350 nm at 900 °C.

7.4.6 Distance Between the Nozzle and Substrate

Since the spraying process from the nozzle to the substrate follows an angle, it is important to optimize the nozzle–substrate distance for efficient film deposition. Even though a small distance ensures a higher deposition rate, the deposition rates vary for each pyrolysis set-up and is determined by the type of atomizer used and the size of the substrate. Charojrochkul et al. [52] have investigated the nozzle–substrate distance and inferred that for a uniform coating, the distance has to be in the range of 3–8 cm. However, this range might depend on the experimental set-up and may vary for other systems. The effect of the nozzle–substrate distance on the deposited film microstructure has been studied extensively. If the distance is short, the substrate is completely covered by an almost dense layer of the film [53]. When the distance is farther, particles will get dispersed on the surface and form powder. As the distance increases further, only a small percentage of droplets reach the surface of the substrate, thereby reducing the spray efficiency [54].

7.5 MATERIALS AND APPLICATIONS

A wide variety of materials is deposited by the SPD technique, which depends on the atomization process for controlling droplet size and distribution on substrates and precursor starting materials [17]. SPD is an easy synthesis process that exhibits a wide range of applications in solar cells, photodiodes, water splitting, energy storage, biomedical devices, etc. [55]. Figure 7.6 depicts the road map of the history of SPD techniques in numerous fields.

Broadly, the materials deposited using SPD technique can be classified as (i) chalcogenides, (ii) metal oxides/ceramics, and (iii) metal and metal alloys such as gold (Au), silver (Ag), and Au–Ag metal alloy. Table 7.1 shows a concise review of different materials deposited using spray pyrolysis.

7.5.1 Chalcogenides

Chalcogenides are reported to be the primary category of materials that have been deposited using SPD [56]. Cadmium-based chalcogenides such as CdS film deposited through this process are widely used in solar cells, gas sensor, and anti-reflection coating applications [21, 65]. Cd precursors such as acetate, formate, chloride, and nitrate, and sulfur precursors such as ammonium thiocyanate, thioacetic acid, thiourea, allylthiourea, 2-thiazoline-2-thiol, and N,N-dimethylthiourea, among others, are commonly used for SPD [56]. A good quality CdS film is obtained by mixing

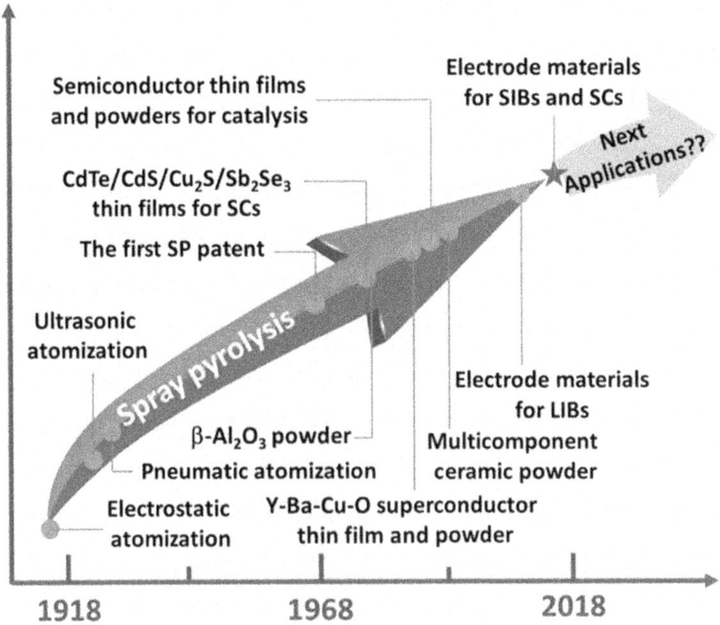

FIGURE 7.6 Growth of the SPD technique in different fields [55].

$CdCl_2$ and thiourea solution, and deposition occurs at a decomposition temperature of 210 °C. The CdS deposited films contain high resistance properties and thus are used for electrophotographic applications [56]. Similarly, CdSe and CdTe films are also obtained through SPD methods [18, 34]. $CdCl_2$ and N, N dimethylselenourea are precursors required for producing CdSe thin films. For CdTe films, the precursors used are $CdCl_2$ and TeO_2 in the presence of reducing agent hydrazine hydride at certain operating parameters such as 300–400 °C, pH at 11.2 with 1–1.5 ml/min of flow rate on a substrate. Deposited CdTe films are obtained with and without an electric field and utilized for photovoltaics applications [34]. Apart from these materials, other cadmium-based materials such as $Cd_xZn_{1-x}S$, $CdSb_2S_4$, K-Sb-S, $Cd_{0.8}Al_{0.2}S$ are also deposited by the SPD process [59]. Thin films based on copper-based chalcogenides are also prepared using the SPD process. For deposition of various phases of Cu_xS, where $1.75 \leq x \leq 2$, the precursors used are cuprous chloride and thiourea at temperatures in the range of 240–320 °C [58]. Thin films of ternary [55] ($CuInS_2$, $CuInSe_2$ etc.) and quaternary [60] (Cu_2ZnSnS_4/Se_4, Cu_2CdSnS_4/Se_4, Cu_2InSnS_4/Se_4, $CuGaSnS_4/Se_4$, $CuIn_5S_4/Se$) copper-based chalcogenides are deposited using SPD methods. Among these materials, a potential absorber material – Cu_2ZnSnS_4 – is one of the heterojunction thin film solar cells. It has solar cell characteristics, that is optimum bandgap in the range of 1.4–1.5 eV and large absorption coefficient [57]. Figure 7.7 shows the FESEM images of Cu_2ZnSnS_4 thin films prepared at various concentrations using SPD methods [60]. The film micrographs represent the uniform coverage of the whole substrate. Also, an increase in average grain size is observed

TABLE 7.1
Different types of properties affected by different types of defects occurring in thin films

S. No.	Type of Materials	Methods for Deposition	Parameters	Applications
(i)	**Chalcogenides**			
	CdS [56]	SPD	Temp: 320–380 ° C	Electrophotographic
	CdSe [57]	SPD	Temp: 200–300 ° C	Solar thermal absorber
	CdTe [34]	Spray pyrolysis with and without electric field	Temp: 300–400 ° C, flow rate: 1–1.5 ml/min, pH:11.2	Photovoltaics
	Cu_xS [58] (Cu_2S, $Cu_{1.95}S$, $Cu_{1.8}S$, $Cu_{1.75}S$, CuS), etc.	USP Frequency 100 kHz	Temp: 240–320 ° C; nozzle to substrate distance: 36 cm; flow rate: 5 cc/min	Heterojunction solar cell
	$ZnIn_2S_4$ [59]			Hydrogen production
	Cu_2ZnSnS_4 [60]			Hydrogen production
(ii)	**Metal oxides**			
	SnO_2 [61]	SPD	Temp: 340 ° C	Supercapacitor
	TiO_2 [16, 62]	USP	Temp: 300–500 ° C,	PSSC, photoconductor
		SPD	Temp: 250 ° C, nozzle to substrate distance: 20 cm	Protective layer for SS
	ZnO [12]	Ultrasonic spray pyrolysis	Temp: 400–550 °C, flow rate: 2 ml/min, nozzle–substrate distance: 5 cm	Laser, supercapacitor
	MgO [19]	USP	Temp: – 580–680 °C, nozzle–substrate distance: 10 cm, nebulizer power 60 W	Buffer layer for successive growth of ferroelectric materials
	NiO [18]		Temp: 350 °C	
	RuO_2 [63]			Supercapacitor
(iii)	**Alloy**			
	Au [64]	Spray pyrolysis	Temp: 400 °C, nozzle–substrate distance: 22.0 cm, spray rate: 4.0 ml/min, N_2 as carrier gas at 1.5 kgf/cm^2 pressure	Photocatalyst, photovoltaics, lasers, solar cells, etc.
	Ag [10]	Spray pyrolysis, ESP	Temp: 450 °C, flow rate: 8 ml/min, spray rate: 3 ml/min, pressure 2 kgf/cm^2	Optical study
	Au–Ag [10]	Spray pyrolysis, ESP	Temp: 450 °C, flow rate: 8 ml/min, spray rate: 3 ml/min, Pressure 2 kgf/cm^2	Optical study and solar cell

FIGURE 7.7 SEM images of thin films Cu_2ZnSnS_4 prepared for various copper concentrations: C1 = 0.15 M, C2 = 0.20 M, and C3 = 0.25 M. Inset shows the cross-sectional view of Cu_2ZnSnS_4 thin films [60].

of values 142 nm (for 0.15 M), 205 nm (for 0.20 M), and 330 nm (for 0.25 M) as Cu concentration increases. Apart from cadmium and copper chalcogenides, other element/metal chalcogenides such as ZnTe, In_2S_3, Bi_2S_3, Sb_2S_3, Sn_2S_3, and so on are also deposited by the spray pyrolysis method [66, 67].

7.5.2 Metal Oxides/Ceramics

Several metal oxides, such as ZnO, MgO, SnO_2, and TiO_2, are deposited using SPD methods. Tin oxide (SnO_2) is deposited by using $SnCl_4$ solvent [61]. The preferred crystal plane orientation of SnO_2 is formed through a process of nuclei formation, crystal growth, and further growth occurring perpendicular to the substrate. SnO_2 shows good electrical and optical properties of electrical resistivity of 33×10^{-4} Ωcm, 70–75% optical transmittance value. Fluorine-doped SnO_2 shows optimum quality of electrical conductivity and optical transmittance [33]. The two phases of TiO_2 such as anatase and rutile phase are formed at temperatures of 300–500 °C and have a direct bandgap of 3.08 eV and an indirect bandgap 3.65 eV [62]. It shows high photoconductivity. The precursor of titanium acetylacetonate is nebulized ultrasonically in ethanol and deposited at 380 °C in argon career gas of 3.5 ml/min on silica substrate. Further anatase to rutile phase is converted after annealing in oxygen at a higher temperature [16]. High-quality uniform and pinhole-free pristine and doped ZnO films are obtained by the USP method [20]. A bandgap of 3.0–3.2 eV lies for ZnO, and it exhibits luminescence behaviour at RT in the blue-green region. The precursor solution of $ZnCl_2$ was maintained at pH 6 after the addition of HCl acid and ultrasonically deposited at a temperature of 400–550 °C on the substrate [38]. Various parameters viz. precursors, doping concentration, and deposition time influence the optical, electrical, and morphological characteristics of ZnO thin films [12, 15]. Figure 7.8 shows an enhanced deposition duration of nanostructured thin films of ZnO. An aerosol mixture is formed using acetylacetonate of magnesium in an isopropanol medium and spray pyrolysis deposited at 300–550 °C, annealed at 1000 °C to form MgO [19]. Other binary metal oxides such as WO_3, MoO_3, NiO, Bi_2O_3, Cr_2O_3, Al_2O_3, Fe_2O_3, RuO_2, ZrO_2 are also been synthesized by spray pyrolysis techniques for thin-film transistor applications [63, 65, 68]. Apart from these, metallic spinel-type

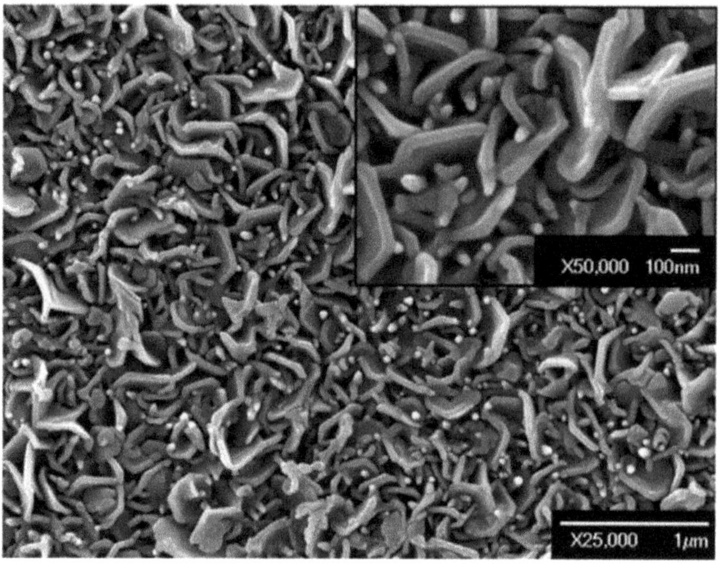

FIGURE 7.8 SEM images of ZnO film synthesized by USP method [12].

oxides [69–71] such as Co_3O_4, $Ni_2Co_2O_4$, MCo_2O_4 (M = Mn, Fe, Cu, Zn), $LiCoO_2$, Pb (Zr, Ti)O_3 (PZT), $KTiOPO_4$, and superconducting oxide thin films such as Hg-Ba-Ca-CuO and Y-Ba-Cu-O are also synthesized by spray pyrolysis techniques [72, 73].

7.5.3 Metal and Metal Alloys Such as Gold (Au), Silver (Ag), and Au–Ag Metal Alloy

For deposition of Au thin films, the precursor solution used is chloroauric acid ($HAuCl_4$). $HAuCl_4$ is dissolved in DI water and conditions for glass substrate deposition at 400 °C with a rate of 4.0 ml/min, nozzle–substrate distance 22.0 cm, and nitrogen career gas at 1.5 kgf/cm^2 pressure [64]. The Au thin film is deposited with the application of a high-voltage electric field (0.5–2 kV) generated between the nozzle and metal electrode below the substrate, and it showed good quality films. The Ag thin film is deposited on the glass substrate as $AgNO_3$ precursor solution and varying the concentration of the solution from 5ml to 10ml. The deposition temperature is 450 °C with carrier gas flow rate 8.0 L/min at 2 kgf/cm^2. Also, the deposition is being carried out by application of electric field [10]. The metal alloy of Au-Ag is obtained by a mixed solution of chloroauric acid ($HAuCl_4$) and $AgNO_3$ at varying molar concentrations [10].

7.6 APPLICATIONS

Spray pyrolysis deposited materials show wide applications in energy storage [31, 55, 74], production [75, 76], solar cell [43], display system [27], and gas sensor [14]. An overview of applications relevant to the current industrial scenario is being briefed here.

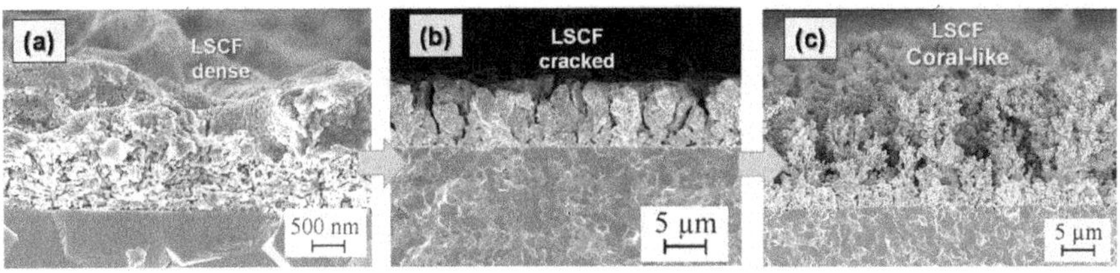

FIGURE 7.9 SEM images of solid oxide fuel cell of cathode electrode [78].

7.6.1 Energy Storage for Batteries (Lithium-Ion) and Supercapacitors

Spray pyrolysis plays an important role in delivering high-energy storage for batteries and achieving high power density for supercapacitors [77]. Materials utilized for energy storage such as lithium-ion batteries, where $LiCoO_2$, $LiFePO_4$, $Li_4Ti_5O_{12}$ etc. act as cathode and metal oxide and carbon composites as anode, show good performance while being deposited using spray pyrolysis. Similarly, metal oxides [74] such as Fe_2O_3, Co_3O_4, RuO_2 -Co_3O_4, Mn_3O_4, Co_3O_4-Mn_3O_4, V_2O_5, which are deposited using spray pyrolysis, are used for supercapacitor purposes. Also, polymers viz. RuO_2/PEDOT:PSS conductive ultrathin films, deposited using SPD techniques are used for the efficient functioning of supercapacitors.

7.6.2 Energy Production for Solid Oxide Fuel Cell (SOFC)

The cathode electrode for SOFCs is composed of rare-earth perovskites such as $La_{1-x}Sr_xCo_2O_{3-\delta}$ $LaSrMnO_3$, $LaAlO_2$, $PrBaCo_2O_6$, $ZnFe_2O_4$, while the anode consists of Ni-based cermets prepared using spray pyrolysis techniques, which exhibit high efficiency due to their intrinsic microstructural design [78]. Figure 7.9 represents the SEM images of a cathode electrode made up of $La_{0.6}Sr_{0.4}Co_{0.8}Fe_{0.2}O_{3-\delta}$ for SOFC.

7.6.3 Solar Cell

Metal oxides exhibit wide optoelectronic applications, and Figure 7.10 shows the morphology of a transparent conducting oxide. Particularly, in Perovskite solar cells the front contact materials are metal oxide (TiO_2, ZnO, SnO_2 etc.) that acts as electron transfer layers and are deposited through the SPD method [16, 13]. Figure 7.10 shows spray pyrolysis-deposited conducting layer materials in a solar cell [13].

7.6.4 Gas Sensor

Mainly spray pyrolysis-deposited metal oxides are widely used in gas sensor devices where a response was observed due to different stimuli by surrounding gases as observed in Figure 7.11 [79]. The binary and perovskite oxides that are spray pyrolysis deposited have large porosity and morphology control of nanorods, nanowires, nanotubes and nanosphere dimensions. Also, the heterostructure of the core-shell structure shows good gas sensitivity [14].

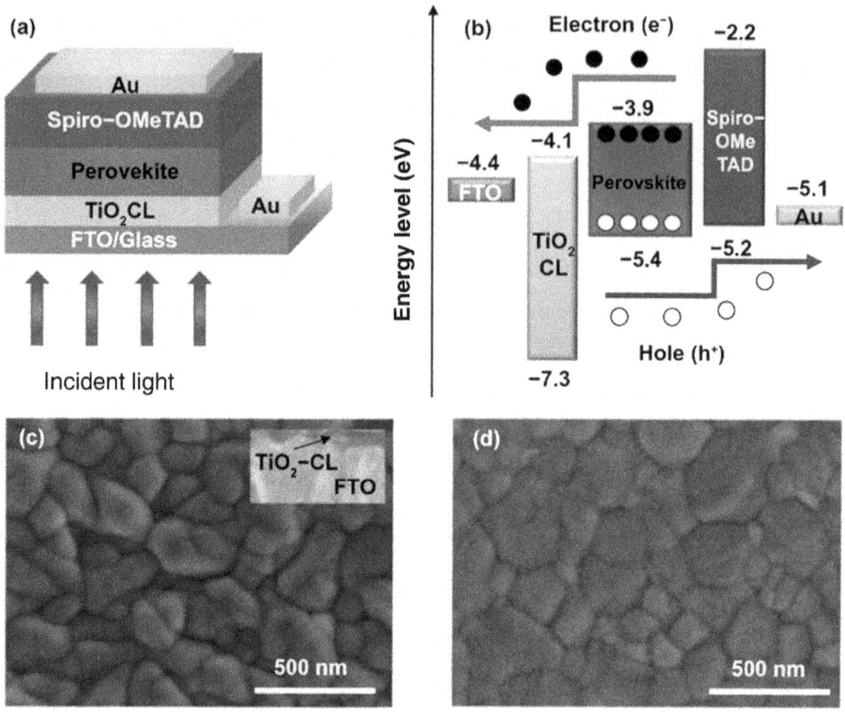

FIGURE 7.10 (a) Single-junction planar perovskite solar cell of the schematic diagram and (b) corresponding energy levels. (c) Top-view SEM image of spray pyrolysis-deposited TiO_2 compact layer and cross-sectional view shown in inset. (d) Perovskite film deposited on TiO_2-CL/FTO substrate [16].

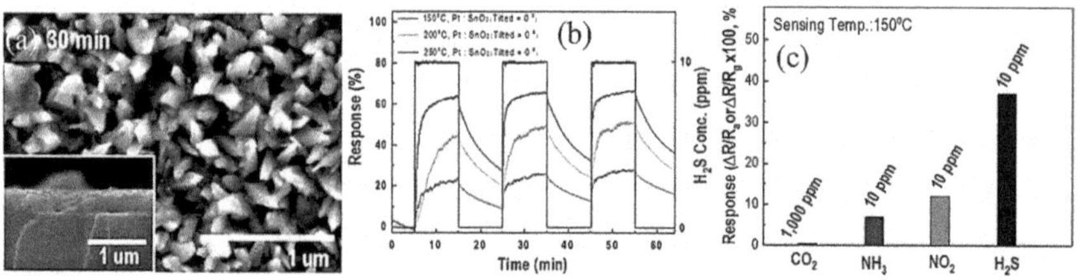

FIGURE 7.11 (a) Gas sensor studies of Pt-doped SnO_2 nanorods of SEM image, (b) H_2S gas responses at operating temperatures, and (c) selectivity towards 10 ppm various gases [79].

7.7 SUMMARY

Spray pyrolysis can be considered a simple process that is economically viable for large-area thin film deposition. Different materials ranging from metal oxides to alloys have been deposited using numerous SPD techniques. Thin film quality depends on optimum process parameters viz. precursor solution, substrate temperature, flow rate, deposition rate, additive for morphology control, droplet size of solution, and distance between nozzle and substrate. The deposited thin films show large applications in solar cells, gas sensing, energy storage batteries, and supercapacitors. These films are considered highly viable for industrial applications, especially in large-area energy storage and conversion devices. However, challenges exist in improving the film quality. Since the deposition process depends on several factors, precise control over the parameters is essential for acquiring quality uniform thin films. SPD methods can generate emissions such as volatile organic compounds and particulate matter, which can contribute to air pollution. Thus, emission control systems of filters and scrubbers are essential to mitigate these impacts, which can lead to complexity in the design.

REFERENCES

1. Venables, J. A. *Introduction to Surface and Thin Film Processes*; Cambridge University Press, 2000.
2. Messier, R. Thin Film Deposition Processes. *MRS Bull*, 1988, 13 (11), 18–21.
3. Budida, J.; Srinivasan, K. Review of Thin Film Deposition and Techniques. In *Materials Today: Proceedings*; Elsevier Ltd, 2023; Vol. 92, pp. 1030–1033.
4. Geyao, L.; Yang, D.; Wanglin, C.; Chengyong, W. Development and Application of Physical Vapor Deposited Coatings for Medical Devices: A Review. In *Procedia CIRP*; Elsevier B.V., 2020; Vol. 89, pp 250–262.
5. Saeed, M.; Alshammari, Y.; Majeed, S. A.; Al-Nasrallah, E. Chemical Vapour Deposition of Graphene – Synthesis, Characterisation, and Applications: A Review. *Molecules. MDPI AG*, September 1, 2020, 25 (17), 3856.
6. Mooney, J. B.; Radding, S. B. Spray Pyrolysis Processing. *Annual Review of Materials Science*, 1982, 12, 81–101.
7. Gunavathy, K. V.; Tamilarasan, K.; Rangasami, C.; Arulanantham, A. M. S. A Review on Growth Optimization of Spray Pyrolyzed Cu_2ZnSnS_4 Chalcogenide Absorber Thin Film. *Int J Energy Res*, 2019, 43 (14), 7716–7754.
8. Maho, A.; Nayak, S.; Gillissen, F.; Cloots, R.; Rougier, A. Film Deposition of Electrochromic Metal Oxides through Spray Coating: A Descriptive Review. *Coatings*, 2023, 13 (11).
9. Annu, A.; Bhattacharya, B.; Singh, P. K.; Shukla, P. K.; Rhee, H. W. Carbon Nanotube Using Spray Pyrolysis: Recent Scenario. *J Alloys Compd*, 2017, 691, 970–982.
10. Kumar, N.; Alam, F.; Dutta, V. Deposition of Ag and Au-Ag Alloy Nanoparticle Films by Spray Pyrolysis Technique with Tuned Plasmonic Properties. *J Alloys Compd*, 2014, 585, 312–317.
11. Workie, A. B.; Ningsih, H. S.; Shih, S. J. An Comprehensive Review on the Spray Pyrolysis Technique: Historical Context, Operational Factors, Classifications, and Product Applications. *Journal of Analytical and Applied Pyrolysis*, Elsevier B.V. March 1, 2023, 170, 105915.
12. Kenanakis, G.; Katsarakis, N.; Koudoumas, E. Influence of Precursor Type, Deposition Time and Doping Concentration on the Morphological, Electrical and Optical Properties of ZnO and ZnO:Al Thin Films Grown by Ultrasonic Spray Pyrolysis. *Thin Solid Films*, 2014, 555, 62–67.
13. Shkir, M.; Chandekar, K. V.; Khan, A.; El-Toni, A. M.; Ashraf, I. M.; Benghanem, M.; Adil, S. F.; Ansari, A. A.; Ghaithan, H.; AlFaify, S. Structural, Morphological, Vibrational, Optical, and Nonlinear Characteristics of Spray Pyrolyzed CdS Thin Films: Effect of Gd Doping Content. *Mater Chem Phys*, 2020, 255.
14. Sriram, S. R.; Parne, S. R.; Pothukanuri, N.; Edla, D. R. Prospects of Spray Pyrolysis Technique for Gas Sensor Applications – A Comprehensive Review. *J Anal Appl Pyrolysis*, 2022, 164 (April), 105527.

15. Rahemi Ardekani, S.; Sabour Rouh Aghdam, A.; Nazari, M.; Bayat, A.; Yazdani, E.; Saievar-Iranizad, E. A Comprehensive Review on Ultrasonic Spray Pyrolysis Technique: Mechanism, Main Parameters and Applications in Condensed Matter. *J Anal Appl Pyrolysis*, 2019, 141 (June), 104631.
16. Shahiduzzaman, M.; Hossain, M. I.; Visal, S.; Kaneko, T.; Qarony, W.; Umezu, S.; Tomita, K.; Iwamori, S.; Knipp, D.; Tsang, Y. H.; et al. Spray Pyrolyzed TiO_2 Embedded Multi-Layer Front Contact Design for High-Efficiency Perovskite Solar Cells. *Nanomicro Lett*, 2021, 13 (1), 1–17.
17. Perednis, D.; Gauckler, L. J. Thin Film Deposition Using Spray Pyrolysis. *J Electroceram*, 2005, 14 (2), 103–111.
18. Patil, P. S. Versatility of Chemical Spray Pyrolysis Technique. *Mater Chem Phys*, 1999, 59 (3), 185–198.
19. Bian, J. M.; Li, X. M.; Chen, T. L.; Gao, X. D.; Yu, W. D. Preparation of High Quality MgO Thin Films by Ultrasonic Spray Pyrolysis. *Appl Surf Sci*, 2004, 228 (1–4), 297–301.
20. Alver, U.; Kilinç, T.; Bacaksiz, E.; Nezir, S. Temperature Dependence of ZnO Rods Produced by Ultrasonic Spray Pyrolysis Method. *Mater Chem Phys*, 2007, 106 (2–3), 227–230.
21. Firat, Y. E.; Yildirim, H.; Erturk, K.; Peksoz, A. Ultrasonic Spray Pyrolysis Deposited Copper Sulphide Thin Films for Solar Cell Applications. *Scanning*, 2017, 2017.
22. Zaouk, D.; Zaatar, Y.; Khoury, A.; Llinares, C.; Charles, J.-P.; Bechara, J. Fabrication of Tin Oxide (SnO2) Thin Film by Electrostatic Spray 2 Pyrolysis. *Microelectronic Engineering*, 2000, 51, 627–631.
23. Kemmler, J. A.; Pokhrel, S.; Mädler, L.; Weimar, U.; Barsan, N. Flame Spray Pyrolysis for Sensing at the Nanoscale. *Nanotechnology*, 2013, 24 (44).
24. Teoh, W. Y.; Amal, R.; Mädler, L. Flame Spray Pyrolysis: An Enabling Technology for Nanoparticles Design and Fabrication. *Nanoscale*, 2010, 2 (8), 1324–1347.
25. Tseng, J. C.; Huang, W. C.; Chang, W.; Jeromin, A.; Keller, T. F.; Shen, J.; Chuang, A. C.; Wang, C. C.; Lin, B. H.; Amalia, L.; et al. Deformations of Ti-6Al-4V Additive-Manufacturing-Induced Isotropic and Anisotropic Columnar Structures: In situ Measurements and Underlying Mechanisms. *Addit Manuf*, 2020, 35.
26. Leeuwenburgh, S. C. G.; Wolke, J. G. C.; Schoonman, J.; Jansen, J. A. Influence of Deposition Parameters on Morphological Properties of Biomedical Calcium Phosphate Coatings Prepared Using Electrostatic Spray Deposition. *Thin Solid Films*, 2005, 472 (1–2), 105–113.
27. Jung, D. S.; Park, S. Bin; Kang, Y. C. Design of Particles by Spray Pyrolysis and Recent Progress in Its Application. *Korean Journal of Chemical Engineering*, 2010, 27 (6), 1621–1645.
28. Ukoba, K. O.; Eloka-Eboka, A. C.; Inambao, F. L. Review of Nanostructured NiO Thin Film Deposition Using the Spray Pyrolysis Technique. *Renewable and Sustainable Energy Reviews*, 2018, 82 (July 2017), 2900–2915.
29. Ravichandran, K.; Jabena Begum, N.; Snega, S.; Sakthivel, B. Properties of Sprayed Aluminum-Doped Zinc Oxide Films – A Review. *Materials and Manufacturing Processes*, 2016, 31 (11), 1411–1423.
30. Chen, H.; Mulmudi, H. K.; Tricoli, A. Flame Spray Pyrolysis for the One-Step Fabrication of Transition Metal Oxide Films: Recent Progress in Electrochemical and Photoelectrochemical Water Splitting. *Chinese Chemical Letters*, 2020, 31 (3), 601–604.
31. Pokhrel, S.; Mädler, L. Flame-Made Particles for Sensors, Catalysis, and Energy Storage Applications. *Energy and Fuels*, 2020, 34 (11), 13209–13224.
32. Kang, Y. C.; Park, S. B. A High-Volume Spray Aerosol Generator Producing Small Droplets For Low Pressure Applications. *Journal of Aerosol Science*, 1995, 26 (7), 1131–1138.

33. Korotcenkov, G.; Brinzari, V.; Schwank, J.; DiBattista, M.; Vasiliev, A. Peculiarities of SnO_2 Thin Film Deposition by Spray Pyrolysis for Gas Sensor Application. *Sens Actuators B Chem*, 2001, 77 (1–2), 244–252.
34. Vamsi Krishna, K.; Dutta, V.; Paulson, P. D. Effect of Electric Field on Spray Deposited CdTe Thin Films. *Thin Solid Films*, 2003, 444 (1–2), 17–22.
35. He, Y.; Hamann, T.; Wang, D. Thin Film Photoelectrodes for Solar Water Splitting. *Chem Soc Rev*, 2019, 48 (7), 2182–2215.
36. Zaier, A.; Oum El az, F.; Lakfif, F.; Kabir, A.; Boudjadar, S.; Aida, M. S. Effects of the Substrate Temperature and Solution Molarity on the Structural Opto-Electric Properties of ZnO Thin Films Deposited by Spray Pyrolysis. *Mater Sci Semicond Process*, 2009, 12 (6), 207–211.
37. Lamri Zeggar, M.; Chabane, L.; Aida, M. S.; Attaf, N.; Zebbar, N. Solution Flow Rate Influence on Properties of Copper Oxide Thin Films Deposited by Ultrasonic Spray Pyrolysis. *Mater Sci Semicond Process*, 2015, 30, 645–650.
38. Owoeye, V. A.; Ajenifuja, E.; Adeoye, A. E.; Salau, A. O.; Adewinbi, S. A.; Akindadelo, A. T.; Pelemo, D. A.; Popoola, A. P. I. Effect of Precursor Concentration on Corrosion Resistance and Microstructure of ZnO Thin Films Using Spray Pyrolysis Method. *Sci Afr*, 2022, 15, e01073.
39. Chen, C.; Buysman, A.; Kelder, E.; Schooman, J. Fabrication of $LiCoO_2$ Thin Film Cathodes for Rechargeable Lithium Battery by Electrostatic Spray Pyrolysis. *Solid State Ion*, 1995, 80 (1–2), 1–4.
40. Adeoye Victor, B. Effect of Concentration and Irradiation on the Optical and Structural Properties of ZnO Thin Films Deposited by Spray Pyrolysis Techniques. *Nucl Instrum Methods Phys Res B*, 2017, 413 (May), 57–61.
41. Laine, R. M.; Marchal, J.; Sun, H.; Pan, X. Q. A New $Y3Al_5O1_2$ Phase Produced by Liquid-Feed Flame Spray Pyrolysis (LF-FSP). *Advanced Materials*, 2005, 17 (7), 830–833. h
42. Fortunato, M. E.; Rostam-Abadi, M.; Suslick, K. S. Nanostructured Carbons Prepared by Ultrasonic Spray Pyrolysis. *Chemistry of Materials*, 2010, 22 (5), 1610–1612.
43. Shinde, P. S.; Bhosale, C. H. Properties of Chemical Vapour Deposited Nanocrystalline TiO_2 Thin Films and Their Use in Dye-Sensitized Solar Cells. *J Anal Appl Pyrolysis*, 2008, 82 (1), 83–88.
44. Nyman, M.; Caruso, J.; Hampden-Smith, M. J.; Kodas, T. T. Comparison of Solid-State and Spray-Pyrolysis Synthesis of Yttrium Aluminate Powders. *Journal of the American Ceramic Society*. Blackwell Publishing Inc., 1997, 1231–1238.
45. Skrabalak, S. E.; Suslick, K. S. Porous MoS_2 Synthesized by Ultrasonic Spray Pyrolysis. *J Am Chem Soc*, 2005, 127 (28), 9990–9991. https://doi.org/10.1021/ja051654g.
46. Song, S. A.; Jung, K. Y.; Park, S. Bin. Preparation of Y2O3 Particles by Flame Spray Pyrolysis with Emulsion. *Langmuir*, 2009, 25 (6), 3402–3406.
47. Attouche, H.; Rahmane, S.; Hettal, S.; Kouidri, N. Precursor Nature and Molarities Effect on the Optical, Structural, Morphological, and Electrical Properties of TiO_2 Thin Films Deposited by Spray Pyrolysis. *Optik (Stuttg)*, 2020, 203.
48. García-Sánchez, M. F.; Ortiz, A.; Santana, G.; Bizarro, M.; Peña, J.; Cruz-Gandarilla, F.; Aguilar-Frutis, M. A.; Alonso, J. C. Synthesis and Characterization of Nanostructured Cerium Dioxide Thin Films Deposited by Ultrasonic Spray Pyrolysis. *Journal of the American Ceramic Society*, 2010, 93 (1), 155–160.
49. Stopic´, S. S.; Nedeljkovićánd, J.; Rakočevicrakočevic´, N. Z.; Uskokovic´institute, D. Influence of Additives on the Properties of Spherical Nickel Particles Prepared by Ultrasonic Spray Pyrolysis. *Journal of Materials Research*, 1999, 14 (7), 3059–3065.
50. Tsai, S. C.; Song, Y. L.; Tsai, C. S.; Yang, C. C.; Chiu, W. Y.; Lin, H. M. Ultrasonic Spray Pyrolysis for Nanoparticles Synthesis. *J Mater Sci*, 2004, 39 (11), 3647–3657.
51. Pradita, T.; Shih, S. J.; Aji, B. B.; Sudibyo. Synthesis of MgO Powder from Magnesium Nitrate Using Spray Pyrolysis. In *AIP Conference Proceedings*; American Institute of Physics Inc., 2017; Vol. 1823.

52. Charojrochkul, S.; Choy, K. L.; Steele, B. C. H. Flame Assisted Vapour Deposition of Cathode for Solid Oxide Fuel Cells. 1. Microstructure Control from Processing Parameters. *J Eur Ceram Soc*, 2004, 24 (8), 2515–2526.
53. Neagu, R.; Perednis, D.; Princivalle, A.; Djurado, E. Zirconia Coatings Deposited by Electrostatic Spray Deposition. Influence of the Process Parameters. *Surf Coat Technol*, 2006, 200 (24), 6815–6820.
54. Hamedani, H. A.; Dahmen, K. H.; Li, D.; Peydaye-Saheli, H.; Garmestani, H.; Khaleel, M. Fabrication of Gradient Porous LSM Cathode by Optimizing Deposition Parameters in Ultrasonic Spray Pyrolysis. *Materials Science and Engineering: B*, 2008, 153 (1–3), 1–9.
55. Leng, J.; Wang, Z.; Wang, J.; Wu, H. H.; Yan, G.; Li, X.; Guo, H.; Liu, Y.; Zhang, Q.; Guo, Z. Advances in Nanostructures Fabricated: Via Spray Pyrolysis and Their Applications in Energy Storage and Conversion. *Chem Soc Rev*, 2019, 48 (11), 3015–3072.
56. Chamberlin, R. R.; Skarman, J. S. Chemical Spray Deposition Process for Inorganic Films. *J Electrochem Soc*, 1966, 113 (1), 86.
57. Chung, C. H.; Bob, B.; Lei, B.; Li, S. H.; Hou, W. W.; Yang, Y. Hydrazine Solution-Processed CuIn(Se,S)2 Thin Film Solar Cells: Secondary Phases and Grain Structure. *Solar Energy Materials and Solar Cells*, 2013, 113, 148–152.
58. Madarász, J.; Bombicz, P.; Okuya, M.; Kaneko, S. Thermal Decomposition of Thiourea Complexes of Cu(I), Zn(II), and Sn(II) Chlorides as Precursors for the Spray Pyrolysis Deposition of Sulfide Thin Films. *Solid State Ion*, 2001, 141–142, 439–446.
59. Li, Y.; Wang, K.; Huang, D.; Li, L.; Tao, J.; Ghany, N. A. A.; Jiang, F. $CdxZn_1$-XS/Sb_2Se_3 Thin Film Photocathode for Efficient Solar Water Splitting. *Appl Catal B*, 2021, 286.
60. Bhosale, S. M.; Suryawanshi, M. P.; Kim, J. H.; Moholkar, A. V. Influence of Copper Concentration on Sprayed CZTS Thin Films Deposited at High Temperature. *Ceram Int*, 2015, 41 (7), 8299–8304.
61. Okuya, M.; Kaneko, S.; Hiroshima, K.; Yagi, I.; Murakami, K. Low Temperature Deposition of SnO_2 Thin Films as Transparent Electrodes by Spray Pyrolysis of Tetra- n-Butyltin (IV). *J Eur Ceram Soc*, 2001, 21 (10–11), 2099–2102.
62. Xu, W. W.; Kershaw, R.; Dwight, K.; Wold, A. Preparation and Characterization of TiO_2 Films by a Novel Spray Pyrolysis Method. *Mater Res Bull*, 1990, 25 (11), 1385–1392.
63. Falcony, C.; Aguilar-Frutis, M. A.; García-Hipólito, M. Spray Pyrolysis Technique; High-K Dielectric Films and Luminescent Materials: A Review. *Micromachines*, 2018, 9 (8), 414.
64. Kumar, N.; Komarala, V. K.; Dutta, V. Deposition of Gold Nanoparticle Films Using Spray Pyrolysis Technique: Tunability of SPR Band by Electric Field. *Physica Status Solidi – Rapid Research Letters*, 2012, 6 (9–10), 406–408.
65. Thomas, S. R.; Pattanasattayavong, P.; Anthopoulos, T. D. Solution-Processable Metal Oxide Semiconductors for Thin-Film Transistor Applications. *Chem Soc Rev*, 2013, 42 (16), 6910–6923.
66. Bouroushian, M. *Electrochemistry of Metal Chalcogenides*; Springer Berlin Heidelberg: Berlin, Heidelberg, 2010.
67. Fiedler, C.; Kleinhanns, T.; Garcia, M.; Lee, S.; Calcabrini, M.; Ibáñez, M. *Solution-Processed Inorganic Thermoelectric Materials: Opportunities and Challenges* ▽. *Chemistry of Materials*; American Chemical Society, October 11, 2022; pp. 8471–8489.
68. Adamopoulos, G.; Thomas, S.; Wöbkenberg, P. H.; Bradley, D. D. C.; McLachlan, M. A.; Anthopoulos, T. D. High-Mobility Low-Voltage ZnO and Li-Doped ZnO Transistors Based on ZrO_2 High-k Dielectric Grown by Spray Pyrolysis in Ambient Air. *Advanced Materials*, 2011, 23 (16), 1894–1898.
69. Patil, V. D.; Patil, D. A.; Jadhav, A. L.; Jadhav, S. L.; Kadam, A. V.; Dandwate, S. R.; Shinde, B. R. Spinel Nickel Ferrite ($NiFe_2O_4$) Materials Synthesized via Spray-Pyrolysis for Electrochemical Supercapacitor Application. *Discover Electrochemistry*, 2024, 1 (1).

70. Vozniuk, O.; Tabanelli, T.; Tanchoux, N.; Millet, J. M. M.; Albonetti, S.; Di Renzo, F.; Cavani, F. Mixed-Oxide Catalysts with Spinel Structure for the Valorization of Biomass: The Chemical-Loop Reforming of Bioethanol. *Catalysts*, 2018, 8 (8), 332.
71. Fortunato, G.; Oswald, H. R.; Reller, A. Spinel-Type Oxide Catalysts for Low Temperature CO Oxidation Generated by Use of an Ultrasonic Aerosol Pyrolysis Process. *J Mater Chem*, 2001, 11 (3), 905–911.
72. Singh, H. K.; Saxena, A. K.; Srivastava, O. N. The Formation and Characterization of T1 Doped Hg-Ba-Ca-Cu-O Superconducting Films Synthesised through Spray Pyrolysis; 1996; Vol. 262, pp. 7–12.
73. Hussain, A. A.; Sayer, M. Microstructure and Superconducting Properties of Y-Ba-Cu-O Films Prepared by Chemical Spray Pyrolysis. *Journal of Superconductivity*, 1991, 4 (5), 385–392.
74. Kate, R. S.; Pathan, H. M.; Kalubarme, R.; Kale, B. B.; Deokate, R. J. Spray Pyrolysis: Approaches for Nanostructured Metal Oxide Films in Energy Storage Application. *J Energy Storage*, 2022, 54 (May), 105387.
75. Vijayan, K.; Vijayachamundeeswari, S. P.; Sivaperuman, K.; Ahsan, N.; Logu, T.; Okada, Y. A Review on Advancements, Challenges, and Prospective of Copper and Non-Copper Based Thin-Film Solar Cells Using Facile Spray Pyrolysis Technique. *Solar Energy*, 2022, 234 (402), 81–102.
76. Jia, X.; Kan, Y.; Zhu, X.; Ning, G.; Lu, Y.; Wei, F. Building Flexible $Li_4Ti_5O_{12}$/CNT Lithium-Ion Battery Anodes with Superior Rate Performance and Ultralong Cycling Stability. *Nano Energy*, 2014, 10, 344–352. https://doi.org/10.1016/j.nanoen.2014.10.012.
77. Nie, P.; Xu, G.; Jiang, J.; Dou, H.; Wu, Y.; Zhang, Y.; Wang, J.; Shi, M.; Fu, R.; Zhang, X. Aerosol-Spray Pyrolysis toward Preparation of Nanostructured Materials for Batteries and Supercapacitors. *Small Methods*, 2018, 2 (2), 1–24.
78. dos Santos-Gómez, L.; Zamudio-García, J.; Porras-Vázquez, J. M.; Losilla, E. R.; Marrero-López, D. Recent Progress in Nanostructured Electrodes for Solid Oxide Fuel Cells Deposited by Spray Pyrolysis. *J Power Sources*, 2021, 507 (April).
79. Eom, N. S. A.; Cho, H. B.; Lim, H. R.; Kim, B. S.; Choa, Y. H. Facile Tilted Sputtering Process (TSP) for Enhanced H_2S Gas Response over Selectively Loading Pt Nanoparticles on SnO_2 Thin Films. *Sens Actuators B Chem*, 2019, 300 (January), 127009.

8 Application of Smart Coatings in Aerospace and Oil and Gas Sectors

Alisha Kakkar and Sudesh Kumar

8.1 INTRODUCTION

Essentially, intelligent materials depict a modern generation of materials that surpass traditional operational and framework materials. These materials are referred to as smart ones due to their intrinsic characteristics, including self-detecting, self-flexibility, retention capacity, and multifaceted functionalities. The self-adjusting properties of smart materials hold exceptional value in facilitating their adaptation to changing conditions. In today's context, there is a substantial demand for smart materials across diverse sectors due to their capacity to precisely alter elemental properties in response to external stimuli. These stimuli encompass factors such as heat, pressure, magnetic regions, reagents, electric power, radioactive emission, and hydraulic force. These alterations can manifest in changes in material dimensions and form, stiffness, restraint, and fluidity. These transformations are responsible for endowing smart materials with a multitude of essential functions, contingent upon environmental variations. They are termed "intelligent" materials as well because they are distinguished by their cognitive ability when subjected to ecological changes. The versatility of these materials extends to applications in construction engineering, aviation technology, automobile systems, robotics engineering, aquatic systems, healthcare technology, bioengineering, machine design, and environmental technology [1]. Furthermore, these materials possess the capability to identify defects and cracks, serving as testing equipment, and subsequently, they exhibit self-regeneration abilities, often referred to as the self-repairing effect. It is intriguing to observe that the traveler count will be approximately twofold in 2036; reaching a total of 7.8 billion annually, according to the International Civil Aviation Organization [2]. To fulfill this necessity, the aerospace sector is expected to undertake substantial and continuous efforts. With such a vast number of travelers and the soaring demand for airplanes, there exists a significant necessity for cutting-edge automation to efficiently fulfill these demands while prioritizing safety. Among the various elements and components within the aviation sector, materials stand out as one of utmost significance. Numerous varieties of materials are currently employed within the aviation sector, but smart materials are steadily gaining prominence owing to their distinctive attributes. In aviation technology, where many utilizations are conducted in unconfined areas exposed to different environmental shifts, smart materials have

 DOI: 10.1201/9781003750659-8

garnered significant focus in the aviation sector. Consequently, the need for smart materials, associated frameworks, and devices is significant within the aviation sector. Metal conduits applied in transporting diverse substances within the petroleum and gas sectors frequently endure harsh inside and outside atmospheric situations, culminating in the deterioration of materials through rust. Usually made of steel with carbon content, these conduits are susceptible to erosion attacks brought on by acids inside the conduits and their redox reactions with the environment [3]. A myriad of attempts have been financially supported to contend with this substantial issue within the sector. Furthermore, the corrosion encountered in conduits bears both cost-effective and protection consequences, comprising a quarter of the breakdowns that have occurred in the worldwide petroleum and gas sector. Lately, a lack of adequate oversight, control, and implementation of petroleum and gas conduits has resulted in numerous pipe eruptions [4]. By and large, events involving petroleum and gas conduits are assessed under three distinct groupings: (a) risky liquids, which encompass gas condensates, processed hydrocarbon products, and oil; (b) petroleum gas distribution networks, and (c) gas collection systems, designed to transport gas from pits to distant markets. To resolve this issue, different tactics have been conceived to protect both inside and outside corrosion in conduits. Electrochemical protection and outer shielding layers, including triple-layer polyethylene, bituminous coating, bituminous mastic, paraffin, etc. have been deployed to avert outer corrosion in conduits. Chemical procedures, like the use of anti-corrosive chemicals have reduced internal corrosion, but the majority of the aforementioned protection techniques are costly and demand substantial labor; hence, inexpensive and uncomplicated preventive methods for corrosion necessitate development. Lately, various protective coating techniques have been conceived by scientists across the world to protect against rust and contamination of substandard steel conduits. In recent times, water-repellent coverings have emerged as anti-corrosion agents for application in aquatic environments [5]. While smart materials find applications across a wide range of industries, this chapter discusses their uses in the oil and gas and aviation industries.

8.2 APPLICATION OF SMART COATINGS

8.2.1 Aviation Sector

The aviation field faces significant financial burdens to reduce costs, enhance efficiency, and maintain rigorous safety standards. As a result, the military-industrial and aerospace technology sectors are actively seeking intelligent materials to meet the objectives. These materials must demonstrate consistency, robustness, and compatibility with niche uses. The attributes of airplane uses are provided in Table 8.1.

Table 8.2 provides an overview of diverse intelligent materials employed in the aviation field. Smart materials have found practical applications in both commercial and economic contexts, with technical uses falling into the categories of detectors and drive units. The smart properties of these materials are achieved through alterations in structure, production methods, defect induction, and alterations to microscopic structures. In specific instances, smart materials serve as load-carrying

TABLE 8.1
The necessities, uses, and impacts of aircraft applications

Necessities	Uses	Impacts
Traveler protection	Transport providers	Utilization of flame-resistant materials Rigorous crash testing
Low density	All aviation initiatives and projects	Construction of semi-monocoque structures Utilization of lightweight materials Strong and lightweight
Durability	All aviation initiatives and projects	Rigorous quality management Comprehensive testing for reliable data Validation of design
Weather monitoring	Airplane	Thunderbolt safeguard Corrosion resistance
Erosion (heat and electromagnetic radiation)	Space vehicle	Durability challenges and lifespan assessment under various circumstances Thin layers with outstanding durability

TABLE 8.2
Different intelligent materials employed in the aviation sector

Necessities	Type of Materials	Uses
Development of intelligent materials for rotor adjustment devices	Magnet-sensitive materials	Periodic and dynamic control Result in limited maneuvering capabilities due to size and weight limitations
Control interface integrated into airplane wings	Shape memory alloys	To address constraints in airplane wing structure
Morphing technology	Morphing intelligent materials	For defense planes
Morphing technology	Shape memory alloys	Developing flexural and elongation properties in wing assemblies and unhinged systems
Morphing technology	Piezo material	Developing flexural and elongation properties in wing assemblies and unhinged systems

devices in plane wings and rotary wing blades. Certain smart materials are employed to mitigate oscillations and adapt the shape of rotor propellers.

Boeing, for example, uses intelligent rotor blade technology, which lowers oscillations by approximately 80%, resulting in improved aircraft efficiency. Moreover, different adjustable control surfaces have been manufactured for aircraft wings. Research efforts are ongoing to advance management methods and design methods for smart materials and to optimize the placement of motor and sensing devices. Additional applications involve the capacity to control the aerial elasticity shape of plane wings to diminish drag and enhance operational efficiency, control oscillations in the low-mass satellite frameworks, and monitor structural integrity in both aerospace vehicles and spacecraft. The dielectric and electromechanical materials find utility in crafting aircraft sensors and converters, including gyro and accelerative sensors. Gyroscopes serve the purpose of determining plane pitch, rapid motion, satellite orientation, and missile tracking, whereas accelerometers are deployed for oscillation analysis. Sensors play a crucial role in determining fuel tank levels. For instance, the Boeing 777 boasts dielectric and electromechanical materials within 60 sonar gasoline tanks. Smart ceramic materials are incorporated into aviation components because they can withstand strong oscillations, collision, and thermal fluctuations. They are employed in the fabrication of diverse aviation components, such as gas pipe assemblies, temperature sensors, and jet engines. Alumina, in particular, is employed for crafting plane components known for their resistance to heat stress and elevated temperatures, both prevalent during spark discharge. Additionally, alumina's lightweight properties contribute to cost savings associated with satellite deployments. The piezo motor plays a pivotal role in enhancing adaptability and efficiency by executing control system functions with exceptional flexibility. Moreover, these materials are harnessed to mitigate oscillations and reduce plane interior noise. Ceramic fibers find application as thermal insulators for flame resistance and heat retention in aviation and space vehicles due to their thermal resistance, unheavy nature, and corrosion resistance. They possess other vital attributes, including a high melting point, flexibility, stretching capacity, and inert properties [6]. Furthermore, airplane glass panels are equipped with a clear, conductive ceramic layer infused within the glass to prevent fogging and ice buildup. Shape memory alloys (SMAs) are utilized to surpass the limitations of conventional plane wing structures. Typically, wings optimized for fast motion struggle to perform effectively at slow speeds, and vice versa. However, the incorporation of these alloys into airplane wings holds promise in addressing this issue. SMAs have a bright future in plane wing frameworks, where they can reduce air resistance losses and optimize aerial pace-wing coordination.

8.2.2 Shape Memory Metal Alloys (SMMAs)

As technology advances, new materials are appearing in the aviation industry. Shape memory metal alloys (SMMAs) are notable materials. As implied by the name, these materials can store and regain their prior forms when exposed to magnetic field and thermomechanic conditions. The fundamental principle behind SMMAs is quite simple: the material undergoes deformation when extrinsic pressure is applied, and it returns to its original shape when subjected to heat until it surpasses a particular temperature point. SMMAs find significant utility in the aerospace sector, particularly in the construction of a plane's engine and wing structure. With changes in velocity, the plane's wings transition to a swept position at a larger Mach level and revert to their original configuration at a slower speed. The shape memory has two distinct stages: the primary stage or Austenite and the resulting stage (Martensite form). The primary phase exhibits an elevated temperature range in contrast to the resulting stage. Several heat removal methods have been employed to control the thermal process and increase the restoration rate. SMMAs containing copper are notable for their affordability and effectiveness during fabrication. Copper-based SMMAs exhibit a rubber-like nature as a result of aging while in the resulting stage. SMMAs containing iron serve as binders for single-use applications due to their exceptional cost-effectiveness; but it has also been observed that iron-based SMMAs boast large strength and superb hyperelasticity. Some SMAs exhibit both the shape memory effect and the temperature memory effect, causing the material to flex back and forth under a specific range of temperatures. Shape memory Ni-Ti alloy enjoys widespread use in the mechanical field due to its geological applications. Dimitrios A. Exarchos and colleagues [7] performed a statistical examination involving different SMMAs. They employed these alloys as coverings to assess their capacity to alleviate strain in subsidiary frameworks by subjecting the SMMAs to practical temperatures. The aim was to increase the material stability and bolster the hardness of these structures, thereby enabling them to withstand substantial loads. The analysis revealed that employing a biphasic SMMA coating on the framework could lead to considerable strain relief.

8.2.3 Shape Memory Polymer Materials

Shape memory polymers (SMPs) are elements capable of storing their initial form even after being distorted. When subjected to extrinsic response, polymers can undergo a change in structure, making them highly applicable in the aerospace sector. Polyurethane stands out as the most commonly utilized polymer, while polystyrene and epoxy also serve as notable SMPs. However, these polymers do have limitations, including poor heat conduction and reduced strength compared to different materials. The transition point of the polymer, known as the glass transition temperature, is the point at which it takes on the characteristics of rubber and exhibits reduced rigidity. Electrically sensitive polymers incorporate conductive filler materials and exhibit high sensitivity to electric current. In the aviation context, these SMPs find utility in transforming outer layers for airplanes. In their amorphous condition, they can handle fluid dynamic pressures and accommodate significant distortions. SMPs with carbon fiber strengthening can serve as pivot actuators, facilitating plane wing distortions. SMPs possess the remarkable capability of full strain recovery. Polyurethanes are particularly adept at rebounding from substantial distortions, making them promising candidates for use in airplane bodies. Epoxy-reinforced shape memory materials, on the other hand, are suitable for plane wings due to their capacity for substantial change in structure and extensive elastic deformations. Tong Mu and colleagues [8] have explored the diverse attributes of SMP Composites. While SMPs offer a range of properties like lightweight and compatibility with living systems, they also present several disadvantages. It is for this reason that polymer composites have gained prominence, offering superior load-bearing capability and larger Young's modulus. These polymer composites comprise nanostructured fiber and short and extended fiber strengthening. They can make autonomous configurations with multifaceted applications. In addition to enhancing the structural attributes of polymers, polymer composite materials introduce innovative activation methods for regeneration. The polymer materials that incorporate a disordered actuating agent are interconnected, providing notable advantages in terms of reusability and ease of processing,

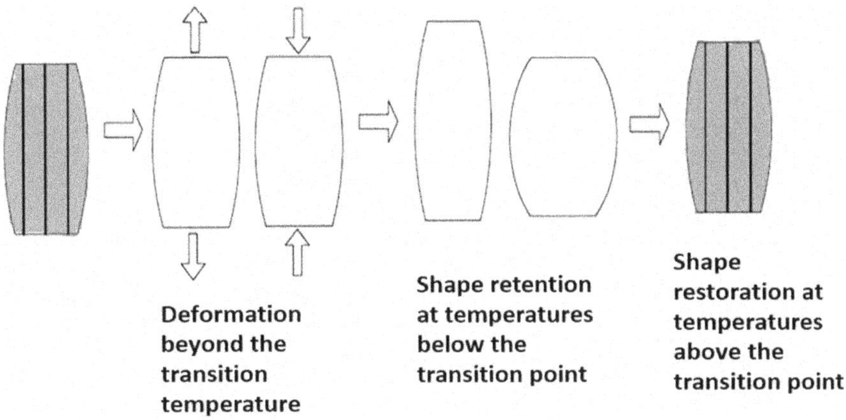

FIGURE 8.1 Representation of shape memory polymers on a large-scale (reproduced from [9]).

as depicted in Figure 8.1. On the other hand, the materials with non-reversible distortions are referred to as chemically interconnected SMPs, offering favorable physical characteristics.

8.3 COMPOSITES MATERIALS

The aviation industry uses intelligent materials extensively because of their low cost, remarkable performance, and simplicity of production. Intelligent composites have the capability to alter their structure when subjected to temperature, strain, or extrinsic forces. A comprehensive overview by Gautam Choubey et al. delves into various composite types and their utility in the aviation sector [10]. For supersonic airplanes like supersonic combustion ramjets and spacecraft re-entry capsules, composites such as polymer composite materials prove invaluable due to their ability to overcome extreme heat conditions. Metallic composite materials employ low-density yet resilient aluminides, offering a large load-bearing efficiency. Ceramic composite materials find use in supersonic combustion ramjets design, as they permit elevated temperatures in the power system, enhancing combustion efficacy. Composite materials serve in space debris detection, monitoring their impact on space satellites. Intelligent carbon-fiber-reinforced polymer plays a crucial role in the space debris sensor. De Simone et al. have introduced a demonstrative prototype for the space debris sensor employing carbon-fiber-reinforced plastic. This sensor can provide early warnings to spacecraft about debris effects [11].

8.3.1 Piezoelectric Materials

Piezoelectric materials (PMs) have experienced significant advancements in the aviation sector. When subjected to external strain, these materials generate electric current and can change forms under the influence of an electrostatic environment as shown in Figure 8.2. Various parts of aircraft, including propulsion systems, wings, and airframe bodies, have incorporated these smart materials. Piezocomposite

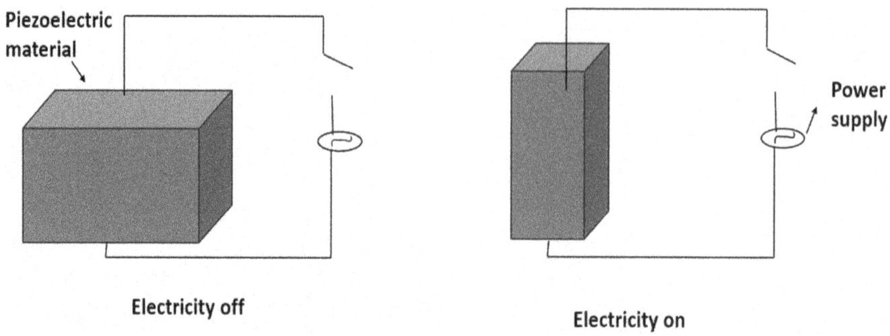

FIGURE 8.2 Structural change of piezoelectric material under the influence of electric energy (reproduced from [17]).

materials, valued for their activation capabilities, reliability, and longevity, have become essential components in the aerospace sector. Integrating the piezo composite materials with carbon-reinforced composite materials has further improved their efficiency, leading to their use in the aerospace industry. Jinqiang Li et al. formulated a kinematic equation for a tailored piezo material, employing PM as an actuating device [12]. Experimental assessments were conducted to evaluate the oscillatory motion of functionalized PM. The mathematical equation was derived from variational methods. Remarkable results can be achieved if the framework of the tailored piezo material plate is improved. PMs have gained popularity in the realm of nanoscopic and microscopic mechanical devices within the aerospace sector. These materials harness energy from external factors to power various devices. Hassan Elahi and coworkers developed a prototype of a nonlinear piezo aerodynamic-elastic harvesting device, operating on the concept of Limit Cycle Oscillations, occurring after reaching the critical flutter speed [13]. The collected energy from the prototype could be utilized to power nanoscopic and microscopic mechanical devices, resulting in a generated power of 0.055 mW from the piezo aerodynamic-elastic harvesting device. Elahi et al. examined the performance of piezo material known as barium titanate in various heat and electric environments while subjecting it to changing loads [14]. Numerical analysis using tools such as Abacus was employed. The study provided a computational forecasting of the impact of thermal and mechanical stress. It was observed that enhancing the resistance led to a decrease in the electric potential generated by barium titanate under varying shock frequencies, ranging from 50 Hz to 250 Hz. However, after some time, the electric potential increased with a slight increase in mechanical shock, indicating that electric potential depends more on shock than resistance.

8.3.2 Reinforced Carbon-Fiber Composite Materials

Because of their low density, high tensile strength, anti-corrosion properties, and structure flexibility, reinforced carbon-fiber composite materials represent an outstanding material for utilization in the aerospace sector. These materials are predominantly employed in the creation of aircraft design. Although these materials show good electric conductance initially, they slowly diminish due to the persistent strain and continuous extrinsic factors. Consequently, these materials tend to become partially conducting over time. Qi Dong and colleagues conducted an investigation into the destruction characteristics of these materials when subjected to electrical components D and C [15]. Their model indicated that heat generated through resistance was responsible for destructing these materials in part D, while the lightning arc heating effect was responsible for damaging part C. Herath and associates provided insights into the thermodynamic and mechanical characteristics of shape-memory polymer composite materials created from pre-impregnated materials [16]. Their empirical findings demonstrated that the introduction of carbon fibers significantly enhanced the physical properties, making these polymer composites suitable for aerospace due to their exceptional attributes.

8.4 OIL AND GAS INDUSTRIES

8.4.1 Self-Cleaning

"Self-cleaning," a characteristic typically encountered in high water-repellent coatings, denotes the capacity of surfaces to autonomously remove unwanted substances like dust and contaminants independently. Auto-cleaning surfaces can be achieved through the following approaches: first, by incorporating photoactive supplements like titanium dioxide, which decomposes unwanted organic particles upon exposure to illumination; second, by creating a water-repellent coating [18]. Furthermore, combining grainy surfaces with low surface adhesion materials can augment the highly water-resistant nature of materials for applications in self-cleaning. Bai and Zhang produced an auto-cleaning water-repellent surface made of graphene-based nickel composite on a 304 metal sheet using an electrodeposition method [19]. In this process, stainless steel plates served as negative electrodes after being finished with carborundum sandpaper, while ultrasound-treated nickel panels were utilized as positive electrodes. The auto-cleaning effectiveness of the synthesized coating was assessed by applying graphitic particles as impurities onto its surface. The weak surface bonding of the synthetic material for coating allowed water to completely remove these contaminants without the need for a surface with a slope. In a different research study, an autocleaning water-repellent Cu-Zn layer on a metal surface was created via the electrodeposition technique [20]. Ash microspheres with diameters of 130 μm, 250 μm, and 90 μm were distributed as contaminants on the material's surface, which was angled at a about 6.5°, to assess the material's capacity for automatic cleaning. Subsequently, liquid drops were discharged onto the polluted area. The mobile liquid droplets washed the surface by displacing the hollow microspheres in their route and the movable drops of water were not hindered by the hollow microspheres. This remarkable property of self-cleaning was attributed to the multi-level surface arrangement, enabling it to trap sufficient air and efficiently prevent interaction between the hollow microspheres and the surface.

8.4.2 De-Icing

Oil leakage in tubes can arise due to the accumulation of ice in extremely cold conditions. Ice has the potential to solidify parts, including regulators, or collect within the tubes [21]. Formerly, incidents involving blockages caused by substances like wax could result in expenses of up to $30,000,000 for a one-time occurrence. In seabed pipelines, such incidents could necessitate scrapping the complete unit, leading to expenses as high as $100,000,000. The formation and accumulation of ice are natural and unavoidable incident that takes place during freezing conditions [22]. Ice buildup takes place on the pipeline exterior utilized for shipping oil and various petrochemical substances, influencing performance within the petroleum industry [23–26]. Now-a-days, several methods are used to reduce or prevent ice buildup. These ice control methods can be categorized as either inactive or proactive measures [27]. The ice control methods encompass heat, mechanical, electric, and ice inhibitors, but these methods are energy-intensive, costly, and often involve toxic substances, in spite of being employed in commercial uses. Conversely, passive icing prevention methods involve the use of anti-icing coatings to protect ice buildup. These methods are cost-effective, environmentally friendly, and easy to implement. Consequently, the adoption of ultra-water-repellent coatings has garnered significant focus lately, both for industrial and scientific purposes [28, 29].

8.4.3 Biofouling Prevention

The buildup of bacteria, aquatic plants, and microorganisms on submerged surfaces is known as biofouling [30]. When a surface, whether natural or man-made, is submerged in marine water, it becomes susceptible to microbial adhesion. Aquatic fouling has evolved into a worldwide concern affecting various sectors, including marine, petroleum, and waterborne industries like submerged structures, ships, pipeline grids, and salt removal plants. The yearly financial toll imposed on worldwide sectors in combating fouling reaches billions of capital. Therefore, it is imperative to devise technologies to counteract the biochemical processes of microbes on structures. Recent studies have unveiled the connection between antimicrobial efficacy and surface wetness [31]. Consequently, extremely water-repellent surfaces have become a viable solution to address fouling problems. These surfaces possess man-made micro-nano laminated structures with poor wetting properties that efficiently reduce the microorganism's adherence to the material's surface [32]. Nanosilver particles are renowned for their resistance to microbes when used in the synthesis of extreme water-repellent surfaces. Feng and other coworkers developed a zinc surface with super-hydrophobicity with enhanced biofouling prevention efficacy over steel grade X65 via electrodeposition and stearic acid treatment [33]. The antimicrobial efficacy of the zinc nanoparticulate covering in turbid water and soil was assessed qualitatively. The material was submerged in turbid water and stayed unpolluted, devoid of any impurities. The anti-fouling attributes of the zinc nanoparticulate covering in turbid water and soil were due to the weakened surface bonding caused by stearic acid treatment and the micro-nano-level coating. Rajab and coworkers engineered ultra-water-repellent coatings over stainless metal using a short-pulse laser process to increase its fouling control characteristics [34]. Bacterial experiments showed that the irregularity in the surfaces caused by the laser treatment decreased the concentration of *Escherichia coli* on the surfaces.

8.4.4 Corrosion Resistant

Xiang and coworkers developed a durable water-repellent surface with outstanding resilience to rust via the electrodeposition method [35]. Initially, a nickel surface with a micro-nano pattern was deposited onto the iron substance. Subsequently, Cr^{3+} covering was electrodeposited onto the previously synthesized nickel layer to enhance its ability to resist corrosion, achieving a corrosion rate of lower than 2.25 × 10–8 A/cm². A transformation from a texture of fish skin to a conical form was noted beyond 20 seconds of electroplating. The micro-scale strength of the nickel/chromium oxide layer was 754 ± 4 HV, and its surface wettability angle was 167.9 ± 2.4 degrees. Ye and coworkers crafted a water-repellent silica material with aniline oligomer content via electrochemical plating on Q235-grade steel to prevent rust [36]. The corrosion-resistant property stemmed from the synergistic effect of the superhydrophobic properties of the electroactive-matrix-silicon dioxide pre-treatment coating and the good activity of the catalyst of the trimeric aniline molecule. This combination effectively blocked the path of corrosive substances and made it easier for protective layers made of iron (III) oxide and iron (II, III) oxide to form on the surface of Q235 steel. Ma and coworkers [37] engineered a liquid-impregnated biomimetic surface on S45C steel by employing a short-pulse laser process accompanied by chemical treatment. The effectiveness of the synthesized coating was assessed using electrochemical polarization and zero-current voltage testing. From these tests, the corrosion potential and corrosion rate values were determined to be 9.57×10^{-8} A/cm² and −0.308 V, respectively. FAS-17 functionalization led to the development of a lubricating coating with minimal surface wettability angle on the S45C steel. The exceptional surface repellency of the surface was attributed to the formation of a multi-layered and optimal porous micro-nano matrix on the S45C carbon steel achieved via efficient laser ablation.

8.4.5 Shear Stress Reduction

The energy utilization associated with global maritime shipping saw a significant surge escalated to 280 million metric tons in 2001 [38]. Astonishingly, half of this energy was expended in conquering frictional resistance. This same challenge plagues the shipping of crude petroleum and a range of petrochemical products through tubes, where a substantial portion of the power transferred by the pump is consumed in countering the wall friction. Lately, the application of water-repellent surface to pipe sheaths has emerged as a practical strategy for curtailing frictional drag [39, 40]. Existing literature reports that irrespective of the flow pattern, whether laminar or turbulent, the creation of a surface air layer by a water-repellent coating on the inner wall of a pipe leads to a reduction in frictional force and induces a certain degree of wall sliding, primarily because the gas fluidity is approximately 50 times lesser than that of the water. In the study by Taghvaei et al., the resistance minimization impacts of water-repellent coatings featuring a double-layer micro-nanoparticle film on pipe flow were thoroughly examined [41]. This investigation unveiled that the micro-nanostructure-coated surface exhibited more frictional force minimization across a broader range of flow velocities compared to nanoparticle-layered surfaces.

The findings revealed a reduction in frictional force ranging from 77% at a Reynolds number of 4.7×10^4 to 7% at 2.35×10^5, all while maintaining an angle of 158.4°. The study suggested that the method employed and the resulting surface could find application in the manufacturing of water-repellent pipelines, thereby mitigating fall in pressure and minimizing overall costs.

8.4.6 Water–Oil Splitting

Lately, frequent oil leakage incidents and industrial effluent release have led to severe pollution and harm to biotic systems worldwide [42, 43]. To address the challenges posed by oil leakage, scientists have created various strategies including solidifying agents, degradation, on-site incineration, oil spill collectors, dispersing agents, and adsorbents, but these technologies come with some disadvantages, including expensive, inefficient separation, delayed segregation rates, and the potential for secondary pollution [44–49]. Consequently, there is a pressing need for extremely effective and cost-effective methods for addressing oil leakage, and modified membranes and water-repellent coatings have emerged as promising solutions [50, 51]. Wen and coworkers created a beneath-oil water-repellent surface and submerged oil-repelling surface on corrosion-resistant steel sieves for water–oil demulsification using the direct vapor deposition method with polydimethylsiloxane in the aqueous phase [52]. The submerged oil-repelling surface effectively separated viscous oil–water blends after oil pre-soaking. They achieved separation efficiencies exceeding 99% for various high- or low-viscous water–oil blends, and this efficacy was maintained beyond 60 filtration cycles. Liu and coworkers synthesized an extreme water-repellent flower-like layered double hydroxide film on corrosion-resistant steel sieves via on-situ growth for effective separation of water–oil mixtures [53]. The resulting surface demonstrated exceptional performance in separating aqua–chloroform blends, achieving 1800 Lm-2 h-1 of radiance and a high efficacy rate of 98.5% after 50 cycles. The potential of the surface to segregate aqua–oil blends was due to its large penetration stress of 5.39 kPa. Therefore, the synthesized film has the potential to serve as a perfect choice for large-scale industrial water–oil segregation because of its rapid flow, outstanding segregation capacity, good anti-corrosion properties, and high strength.

8.5 CHALLENGES AND FUTURE PROSPECTS

Smart materials represent the latest frontier in human advancement, poised to propel us toward a promising future. In spite of the substantial requirement and future potential of intelligent materials, challenges persist, particularly in terms of quality and affordability. Researchers are tasked with enhancing quality without inflating expenses. Moreover, developing intelligent materials practically hinges on the development of exceptional smart materials, smooth blending into laminated frameworks, the synergy of structural and conductive properties, improved efficiency at a reasonable cost, and advancements in microchip technology, data handling, and sensing systems. This era will bear witness to the dominance of an extensive range of smart materials, particularly in the aviation sector and oil and gas industries. The imperative lies in the development of smart materials capable of shaping aircraft by particular prerequisites and specifications. Furthermore, there's a significant requirement for flexible materials in the realm of shape-shifting technology. Smart materials, with their unique properties, are well-equipped to address the challenges faced by the aviation sector, making research in this domain highly promising. Undoubtedly, the smart materials industry will expand in the coming years, drawing scientists toward addressing aviation technology challenges. Creative research endeavors are anticipated, particularly in the realm of flight control elements and flexible airfoils, which hold the potential to greatly enhance flexibility. The quest for small-sized intelligent materials to minimize oscillation and noise is pressing. Intelligent eco-responsive nano-hydrogels, capable of detecting changes in temperature, acid value, or metabolic level, have been developed and can eject their content in response to such adaptations. Smart nanomaterials, potentially employed in SMMAs, hold great promise, particularly in altering plane wing structures to optimize aerodynamic performance efficiently and economically. The future demands the development of intelligent materials, tailored to the evolving requirements of aircraft technology. The future beckons with prospects like self-repairing systems, shock dampening, auto-recovery abilities, and thermal regulation through nanosized intelligent materials. Nanotechnology is currently attaining prominence across diverse domains. The petroleum sector, for instance, has already embraced nano-scale intelligent materials sensitive to stimuli for various applications, such as expandable rubbery materials in responsive packers and thermal-responsive inflatable displays. The emergence of eco-friendly anti-corrosion agents and protective coatings has captured the attention of researchers. Due to various governmental regulations prohibiting the utilization of harmful substances, the number of papers devoted to the development of effective anti-corrosion materials that are non-hazardous, environmentally friendly, and readily decomposable has increased [54]. The scarcity of precise corrosion records and their connection to operational and procedural circumstances, along with the capability to foresee corrosion locations with exceptional precision, present challenges currently confronted by oil and gas industries. Most of the prevailing prognostic models rely on lab tests, which may or may not faithfully replicate real-world scenarios. The application of durable superhydrophobic intelligent materials is another viable solution. The latest investigations have concentrated on crafting intelligent materials that autonomously react when subjected to harsh conditions [55, 56]. Figure 8.3 depicts the different strategies to prevent corrosion in petroleum industries.

8.6 CONCLUSION

The advancement of intelligent materials constitutes a multidisciplinary area. Intelligent materials represent a modern era material, poised to usher in a transformative era across various domains. The production of nanoscale intelligent materials and comprehension of their operational mechanisms stand to enhance characteristics and applications within different sectors. These intelligent materials embody the aspirations of tomorrow and are poised to significantly elevate living standards. To facilitate a real transformation, cooperative endeavors are imperative, involving scholars, scientists, technicians, and architects. Forthcoming research should

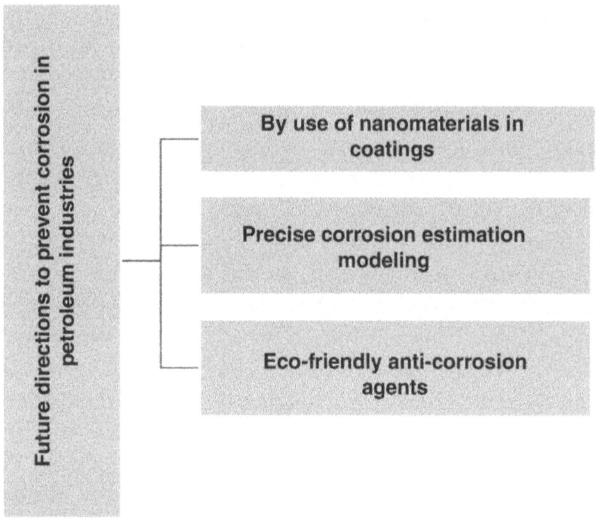

FIGURE 8.3 Future directions to control corrosion in the petroleum sector.

prioritize the innovation of cutting-edge and intelligent water-repellent coverings that integrate inhibitors against corrosion and possess self-repairing abilities. These coatings should be designed to mend their rust-inhibiting, anti-microbial, and auto-repairing properties from extrinsic factors like ultraviolet radiation. It is imperative to create more eco-sustainable and cost-effective water-repellent surfaces, highlighting their extensive range of commercial applications. Latest investigations have predominantly assessed the efficiency of different newly formed water-repellent surfaces over relatively brief periods, typically no longer than two months. Consequently, there is a pressing need to develop water-repellent surfaces capable of sustaining their rust-inhibiting attributes under the influence of harsh degrading environments, particularly in the aquatic regions where pipes are frequently utilized. Additionally, future research should delve into tribological examinations concerning the application of water-repellent surfaces. Previous studies primarily concentrated on metal bases (e.g., X80 and X90 pipelines) for applications in high-density oil and gas mixtures. It is advisable to choose commercial-grade steel bases such as ASME B31.4 and ASTM A106, which are engineered for both high-density and elevated-temperature oil and gas operations. Looking forward, there is an opportunity to innovate and introduce new oil-repellent surfaces to safeguard commercial-scale steel, which has an important role in the shipping of gas and oil under conditions of elevated temperature and pressure.

ACKNOWLEDGMENTS

It is our privilege to express sincere appreciation to the Department of Chemistry, Banasthali Vidyapith, Rajasthan, for providing the laboratory and internet facilities essential for this work. We are also grateful to the reviewers for their valuable guidance, which greatly contributed to the successful publication of this study.

REFERENCES

1. Bashir, A. M. (2017). Smart materials and their applications in civil engineering: An overview. *International Journal of Civil Engineering and Construction Science*, 4, 11–20.
2. Matthew, P. (2018). *Current and Emerging Trends in the Aerospace Sector: How Shifting Priorities and Developing Technologies Are Shaping the Industry Today and into the Future*. SNC-Lavalin'sAtkins. www.atkinsglobal.com//media/Files/A/Atkins-Corporate/aviation-trends-white-paper-digital. pdf.
3. Askari, M., Aliofkhazraei, M., & Afroukhteh, S. (2019). A comprehensive review on internal corrosion and cracking of oil and gas Pipelines. *J. Nat. Gas Sci. Eng.*, 711, 102971.
4. United States Department of Transportation. (2019). *Pipeline Incident 20 Years Trends*. https://www.phmsa.dot.gov/data-and-statistics/pipeline/pipeline-incident-20-yea r-trends.
5. Jurak, S. F., Jurak, E. F., Uddin, M. N., & Asmatulu, R. (2020). Functional superhydrophobic coating systems for possible corrosion mitigation. *Int. J. Autom. Technol.*, 14, 148–158.
6. Basheer, A. A. (2020). Advances in the smart materials applications in the aerospace industries. *Aircraft Engineering and Aerospace Technology*, 92(7), 1027–1035. https://doi.org/10.1108/AEAT-02-2020-0040.
7. Exarchos, D. A., Dalla, P. T., Tragazikis, I. K., Dassios, K. G., Zafeiropoulos, N. E., Karabela, M. M.,. . . & Matikas, T. E. (2018). Development and characterization of high-performance shape memory alloy coatings for structural aerospace applications. *Materials*, 11(5), 832. https://doi.org/10.3390/ma11050832.
8. Mu, T., Liu, L., Lan, X., Liu, Y., & Leng, J. (2018). Shape memory polymers for composites. *Composites Science and Technology*, 160, 169–198.
9. Sharma, K., & Srinivas, G. (2020). Flying smart: Smart materials used in aviation industry. *Materials Today: Proceedings*, 27, 244–250.
10. Choubey, G., Suneetha, L., & Pandey, K. M. (2018). Composite materials used in Scramjet-A review. *Materials Today: Proceedings*, 5(1), 1321–1326. https://doi.org/10.1016/j.matpr.2017.11.217.
11. De Simone, M. E., Andreades, C., Hilmi, A. M., Meo, M., & Ciampa, F. (2019). Proof of concept for a smart composite orbital debris detector. *Acta Astronautica*, 160, 499–508. https://doi.org/10.1016/j.actaastro.2019.02.024.
12. Li, J., Xue, Y., Li, F., & Narita, Y. (2019). Active vibration control of functionally graded piezoelectric material plate. *Composite Structures*, 207, 509–518. https://doi.org/10.1016/j.compstruct.2018.09.053.
13. Elahi, H., Eugeni, M., & Gaudenzi, P. (2019). Design and performance evaluation of a piezoelectric aeroelastic energy harvester based on the limit cycle oscillation phenomenon. *Acta Astronautica*, 157, 233–240. https://doi.org/10.1016/j.actaastro.2018.12.044.
14. Elahi, H., Eugeni, M., Gaudenzi, P., Qayyum, F., Swati, R. F., & Khan, H. M. (2018). Response of piezoelectric materials on thermomechanical shocking and electrical shocking for aerospace applications. *Microsystem Technologies*, 24, 3791–3798.
15. Dong, Q., Wan, G., Guo, Y., Zhang, L., Wei, X., Yi, X., & Jia, Y. (2019). Damage analysis of carbon fiber composites exposed to combined lightning current components D and C. *Composites Science and Technology*, 179, 1–9.
16. Herath, H. M. C. M., Epaarachchi, J. A., Islam, M. M., Al-Azzawi, W., Leng, J., & Zhang, F. (2018). Structural performance and photothermal recovery of carbon fibre reinforced shape memory polymer. *Composites Science and Technology*, 167, 206–214.
17. Chikker, R. (2017). Design of parameters of underwater electro-acoustic Tonpilz transducers (Doctoral dissertation, Bremen University).
18. Nguyen-Tri, P., Tran, H. N., Plamondon, C. O., Tuduri, L., Vo, D. V. N., Nanda, S., Mishra, A., Chao, H. P., & Bajpai, A. K. (2019). Recent progress in the preparation,

properties and applications of superhydrophobic nano-based coatings and surfaces: A review. *Prog. Org. Coating*, 132, 235–256.
19. Bai, Z., & Zhang, B. (2020). Fabrication of superhydrophobic reduced-graphene oxide/nickel coating with mechanical durability, self-cleaning and anti-corrosion performance. *Nano Mater. Sci.*, 2(2), 151–158.
20. Li, H., Yu, S., Hu, J., & Yin, X. (2019a). Modifier-free fabrication of durable superhydrophobic electrodeposited Cu-Zn coating on steel substrate with selfcleaning, anti-corrosion and anti-scaling properties. *Appl. Surf. Sci.*, 481, 872–882.
21. Xu, H., Bbosa, B., Pereyra, E., Volk, M., & Mannan, M. S. (2018). Oil transportation in pipelines with the existence of ice. *J. Loss Prev. Process. Ind.*, 56, 137–146.
22. Wu, X., Silberschmidt, V. V., Hu, Z. T., & Chen, Z. (2019). When superhydrophobic coatings are icephobic: Role of surface topology. *Surf. Coating. Technol.*, 358, 207–214.
23. Barthwal, S., Lee, B., & Lim, S. H. (2019). Fabrication of robust and durable slippery anti-icing coating on textured superhydrophobic aluminum surfaces with infused silicone oil. *Appl. Surf. Sci.*, 496, 143677.
24. Shen, Y., Wang, G., Zhu, C., Tao, J., Lin, Y., Liu, S., Jin, M., & Xie, Y. (2017). Petal shaped nanostructures planted on array micro-patterns for superhydrophobicity and anti-icing applications. *Surf. Coating. Technol.*, 319, 286–293.
25. Golkhou, F., & Haghtalab, A. (2019). Measurement and thermodynamic modeling of carbon dioxide hydrate formation conditions using dry water through hydrophobic nano silica. *J. Nat. Gas Sci. Eng.*, 68, 102906.
26. Lei, S., Wang, F., Fang, X., Ou, J., & Li, W. (2019). Icing behavior of water droplets impinging on cold superhydrophobic surface. *Surf. Coating. Technol.*, 363, 362–368.
27. Latthe, S. S., Sutar, R. S., Bhosale, A. K., Nagappan, S., Ha, C. S., Sadasivuni, K. K., Liu, S., & Xing, R. (2019). Recent developments in air-trapped superhydrophobic and liquid-infused slippery surfaces for anti-icing application. *Prog. Org. Coating*, 137, 105373.
28. Egbo, M., Nasersharifi, Y., & Hwang, G. (2020). Phase-change heat transfer of sintered particle wick in downward facing orientation: Particle size and wick thickness effects. *Int. J. Heat Mass Tran.*, 155, 119840.
29. Liu, Y., Fu, K., Liu, J., Tian, Y., Zhang, H., Wang, R., Zhang, B., Zhang, H., Zhou, F., & Zhang, Q. (2019). Design and preparation of a multi-fluorination organic superhydrophobic coating with high mechanical robustness and icing delay ability. *Appl. Surf. Sci.*, 497, 143663.
30. Sun, K., Yang, H., Xue, W., He, A., Zhu, D., Liu, W., Adeyemi, K., & Cao, Y. (2018). Antibiofouling superhydrophobic surface fabricated by picosecond laser texturing of stainless steel. *Appl. Surf. Sci.*, 436, 263–267.
31. Li, S., Liu, Y., Zheng, Z., Liu, X., Huang, H., Han, Z., & Ren, L. (2019b). Biomimetic robust superhydrophobic stainless-steel surfaces with antimicrobial activity and molecular dynamics simulation. *Chem. Eng. J.*, 372, 852–861.
32. Dahariya, S., Patel, N., Egbo, M. K., Hwang, G., & Betz, A. R. (2019). High-pressure poolboiling heat transfer enhancement mechanism on sintered-particle wick surface. *Front. Mech. Eng.*, 5, 71.
33. Feng, Y., Chen, S., & Frank Cheng, Y. (2018). Stearic acid-modified zinc nano-coatings with superhydrophobicity and enhanced antifouling performance. *Surf. Coating. Technol.*, 340, 55–65.
34. Rajab, F. H., Liauw, C. M., Benson, P. S., Li, L., & Whitehead, K. A. (2018). Picosecond laser treatment production of hierarchical structured stainless steel to reduce bacterial fouling. *Food Bioprod. Process.*, 109, 29–40.
35. Xiang, T., Chen, D., Lv, Z., Yang, Z., Yang, L., & Li, C. (2019). Robust superhydrophobic coating with superior corrosion resistance. *J. Alloys Compd.*, 798, 320–325.

36. Ye, Y., Liu, Z., Liu, W., Zhang, D., Zhao, H., Wang, L., & Li, X. (2018b). Superhydrophobic oligoaniline-containing electroactive silica coating as pre-process coating for corrosion protection of carbon steel. *Chem. Eng. J.*, 348, 940–951.
37. Ma, Q., Wang, W., & Dong, G. (2019). Facile fabrication of biomimetic liquid-infused slippery surface on carbon steel and its self-cleaning, anti-corrosion, anti-frosting and tribological properties. *Colloids Surfaces A Physicochem. Eng. Asp.*, 577, 17–26.
38. Tang, L., Zeng, Z., Wang, G., Liu, E., Li, L., & Xue, Q. (2017). Investigation on superhydrophilic surface with porous structure: Drag reduction or drag increasing. *Surf. Coating. Technol.*, 317, 54–63.
39. Hassani, A. H., & Ghazanfari, M. H. (2017). Improvement of non-aqueous colloidal gas aphron-based drilling fluids properties: Role of hydrophobic nanoparticles. *J. Nat. Gas Sci. Eng.*, 42, 1–12.
40. Chen, H., Di, Q., Ye, F., Gu, C., & Zhang, J. (2016). Numerical simulation of drag reduction effects by hydrophobic nanoparticles adsorption method in water flooding processes. *J. Nat. Gas Sci. Eng.*, 35, 1261–1269.
41. Taghvaei, E., Moosavi, A., Nouri-Borujerdi, A., Daeian, M. A., & Vafaeinejad, S. (2017). Superhydrophobic surfaces with a dual-layer micro- and nanoparticle coating for drag reduction. *Energy*, 125, 1–10.
42. Yu, T., Halouane, F., Mathias, D., Barras, A., Wang, Z., Lv, A., Lu, S., Xu, W., Meziane, D., Tiercelin, N., Szunerits, S., & Boukherroub, R. (2020). Preparation of magnetic, superhydrophobic/superoleophilic polyurethane sponge: Separation of oil/water mixture and demulsification. *Chem. Eng. J.*, 384, 123339.
43. Lee, C. H., Tiwari, B., Zhang, D., & Yap, Y. K. (2017). Water purification: Oil-water separation by nanotechnology and environmental concerns. *Environ. Sci. Nano.*, 4, 514–525.
44. Ullah Khan, I., Othman, M. H. D., Ismail, A. F., Matsuura, T., Hashim, H., Nordin, N. A. H. M., Rahman, M. A., Jaafar, J., & Jilani, A. (2018). Status and improvement of dual-layer hollow fiber membranes via co-extrusion process for gas separation: A review. *J. Nat. Gas Sci. Eng.*, 52, 215–234.
45. Bera, A., & Belhaj, H. (2016). Application of nanotechnology by means of nanoparticles and nanodispersions in oil recovery – a comprehensive review. *J. Nat. Gas Sci. Eng.*, 34, 1284–1309.
46. Ahmad, N. A., Leo, C. P., & Ahmad, A. L. (2018). Amine wetting evaluation on hydrophobic silane-modified polyvinylidene fluoride/silicoaluminophosphate zeolite membrane for membrane gas absorption. *J. Nat. Gas Sci. Eng.*, 58, 115–125.
47. Ge, J., Shi, L. A., Wang, Y. C., Zhao, H. Y., Yao, H.-B., Zhu, Y.-B., Zhang, Y., Zhu, H. W., Wu, H.-A., & Yu, S.-H. (2017). Joule-heated graphene-wrapped sponge enables fast clean-up of viscous crude oil spill. *Nat. Nanotechnol.*, 12, 434.
48. Wang, N., Wang, Y., Shang, B., Wen, P., Peng, B., & Deng, Z. (2018). Bioinspired one-step construction of hierarchical superhydrophobic surfaces for oil/water separation. *J. Colloid Interface Sci.*, 531, 300–310.
49. Mi, H. Y., Jing, X., Xie, H., Huang, H. X., & Turng, L. S. (2018). Magnetically driven superhydrophobic silica sponge decorated with hierarchical cobalt nanoparticles for selective oil absorption and oil/water separation. *Chem. Eng. J.*, 337, 541–551.
50. Zhang, B., Xu, W., Zhu, Q., Sun, Y., & Li, Y. (2019). Mechanically robust superhydrophobic porous anodized AA5083 for marine corrosion protection. *Corrosion Sci.*, 158, 108083.
51. Jiang, Y., Liu, C., Li, Y., & Huang, A. (2019). Stainless-steel-net-supported superhydrophobic COF coating for oil/water separation. *J. Membr. Sci.*, 587, 117177.
52. Wen, N., Miao, X., Yang, X., Long, M., Deng, W., Zhou, Q., & Deng, W. (2018). An alternative fabrication of underoil superhydrophobic or underwater superoleophobic

stainless steel meshes for oil-water separation: Originating from one-step vapor deposition of polydimethylsiloxane. *Sep. Purif. Technol.*, 204, 116–126.
53. Liu, P., Zhang, Y., Liu, S., Zhang, Y., & Qu, L. (2019b). Fabrication of superhydrophobic marigold shape LDH films on stainless steel meshes via in-situ growth for enhanced anti-corrosion and high efficiency oil-water separation. *Appl. Clay Sci.*, 182, 105292.
54. Obot, I. B., Solomon, M. M., Umoren, S. A., Suleiman, R., Elanany, M., Alanazi, N. M., & Sorour, A. A. (2019). Progress in the development of sour corrosion inhibitors: Past, present, and future perspectives. *Journal of Industrial and Engineering Chemistry*, 79, 1–18.
55. Ulaeto, S. B., Rajan, R., Pancrecious, J. K., Rajan, T. P. D., & Pai, B. C. (2017). Developments in smart anticorrosive coatings with multifunctional characteristics. *Progress in Organic Coatings*, 111, 294–314.
56. Sharma, V., Goyat, M. S., Hooda, A., Pandey, J. K., Kumar, A., Gupta, R.,. . . & Bhargav, P. K. (2020). Recent progress in nano-oxides and CNTs based corrosion resistant superhydrophobic coatings: A critical review. *Progress in Organic Coatings*, 140, 105512.

9 Evaluation of the Mechanical Performance of TiN Thin Film Cr and CrN Interlayer

V. S. S. Venkatesh, Sunil Kumar, and Manigandan Sekar

9.1 INTRODUCTION

Tool steel was discovered decades ago, but even today research on it is going on a large scale. This is because there is still room for improvement in its properties to withstand the heavy-duty condition. For example, tool steel is used in tooling industries to manufacture tools, such as press tools, forging dies, die-casting dies, and so on [1, 2]. These tools are further used to shape other materials that may be soft or hard. During the shaping process, direct metal-to-metal contact occurs, and due to this contact, the tool elements face heavy wear, and sometimes they break down also. To reduce the wear of the tool elements, heat treatment is done to improve the surface as well as the volumetric properties of the tool steel [3, 4]. Post-heat treatment helps to improve the wear properties to reduce the wear of the tool elements to some extent, but it does not give a permanent solution. In this regard, much research [5–7] has been conducted and concluded that further surface treatment needs to be done to improve the surface properties. It can be done using plasma nitriding, deposition of thin film, or both, on the heat-treated tool steel.

There are numerous alternatives available to thin films, out of which a popular thin film is TiN (titanium nitride). Because of its ease of deposition and specific properties like wear, mechanical, adhesion, and oxidation properties, it attracts research to solve the problem. Many researchers [8, 9] have suggested that the aforementioned properties of TiN thin film can be improved using the different interlayers.

Research initiatives have used a variety of technologies and coatings. Cr [10, 11], TiN [12, 13], TiN/Al_2O_3/TiCN [14], CrN [15], and multilayer composite coating [16, 17] are a few of the possible coatings that have been investigated. When it comes to improving the surface properties of tool steel, Cr coating has received the most attention in research and is thought to be the most promising option. The first full-length Cr-coated cladding tubes will be made using a reported coating technology that includes specialized PVD processes [18, 19], arc ion plating [20], cathodic arc-physical vapor deposition (CA-PVD) [21], vacuum-arc evaporation methods [14],

DOI: 10.1201/9781003750659-9

sputtering deposition [22], cold spraying [23], laser spraying [24], and more, according to Xiao et al. [25]. The described research also places a significant emphasis on the coatings' attributes, including their microstructural features, adhesion qualities, high-temperature oxidation behavior, water-quenching behavior, and corrosion resistance.

In this study, the effect of interlayer on mechanical, structural, and adhesion properties of TiN film was investigated. First, DAC-10 tool steel was heat-treated and then TiN, Cr/TiN, and Cr-(CrN/TiN) film was deposited using the cathodic-arc-evaporation physical vapor deposition method (CAE-PVD). Structural and morphological characteristics were evaluated post-deposition of film XRD, AFM, and SEM respectively. Nanomechanical and scratch tests were conducted on a tribometer machine to determine the mechanical and tribological properties of deposited films. Conclusions have been drawn at the end of this chapter.

9.2 MATERIALS AND METHODOLOGY

9.2.1 Sample Preparation

The present study chose DAC-10 hot work tool steel with a 10 mm × 10 mm × 8 mm thickness for this experiment, and the chemical composition of this steel was elaborated in the literature [26]. Further sample preparation steps followed the literature, which has been explained in detail [27]. The substrate was heat treated at about 1095 °C in an oxygen-free atmosphere. For the quenching procedure, nitrogen gas was delivered into the chamber at a pressure of around 2 MPa. To temper the metal, the chamber pressure was gradually reduced to around 0.13 MPa while being rapidly cooled at a pace of about 93 °C. The nitrided surface of the DAC-10 tool steel had a hardness of approximately 451 HV, which is comparable to roughly 4.4 GPa or roughly 44 HRC, according to the Mitutoyo hardness testing equipment (model: HV-115).

Prior to depositing, the thin films needed to achieve a good surface roughness says 2 μm, the surfaces were painstakingly polished using sandpaper with grit levels of 100, 250, 400, 800, 1000, and 1400. During the polishing procedure, which was completed using a rotating disc polishing machine, a diamond paste with a particle size of 3 μm was applied. The substrates were then thoroughly cleaned with alcohol and dried by blowing air over them.

9.2.2 Film Deposition

A thin layer of chromium was deposited on DAC-10 tool steel substrates using a chromium plating process. Chromium nitride (CrN) deposition was performed utilizing cathodic arc deposition. As the target material, we used chromium with a purity of 99.99%. After carefully cleaning them, the substrates were heated to a negative bias of −1,200 V. We kept the process running at a base pressure of 210–3 Pa. An inert gas environment made up of argon (Ar) and nitrogen (N_2) was used to keep the substrate temperature at 300 °C and the deposition pressure at 6 10–1 Pa. In addition, a −150 V negative bias was used. We changed the gas composition to produce a multilayer covering; detailed information on the deposition procedure can be found elsewhere [28]. Cathodic arc evaporation (CAE-PVD) was used to create thin films of titanium nitride (TiN) by evaporating titanium (Ti) from a target in a combination of argon (Ar) and nitrogen (N_2). At a raised temperature of 350 °C and a controlled working pressure range of 0.1 to 0.2 Pascal (Pa), the deposition procedure took place. With a substrate bias of –70 volts, a coating thickness of around 2 µm was achieved. Hence, three different types of samples were prepared namely TiN, Cr/TiN, and Cr-(CrN/TiN) deposited films.

9.2.3 Structural and Morphological Characterizations

Using a Malvern Panalytical diffractometer (model: Xpert, Netherlands), X-ray diffraction analysis was used to evaluate the coating's composition and crystalline structure. Using CuK radiation with a wavelength of 1.54, a current of 40 mA, and a voltage of 40 kV, this investigation was conducted at a shallow grazing angle of 1.5°. A precise step size of 0.02° was used to detect faint intensity peaks when the X-ray scan advanced at a pace of 2° per minute over a scanning range of 30° to 70°.

Atomic force microscopy (Agilent Technologies, model: 5100, USA) was used to assess the surface properties, specifically the area roughness parameters, of three distinct materials, namely the uncoated 316 LVM, AlCrN/CrN, and a-C:Ag/CrN coatings. With a height resolution of 50 nm and a scanning area of 40 m, the device resonated at a frequency of 250 kHz. The image resolution was set to 512 points per second, and the scanning operation was carried out at a pace of 0.5 lines per second. The scanning probe used for image acquisition in ambient air using intermittent contact mode was a monolithic silicon scanning tip (PPP-NCL, Nano Sensors, Inc., USA). Additionally, a field emission scanning electron microscope (FESEM) (Zeiss Supra, model: VP-55 FEGSEM, Germany) captured SEM pictures.

9.2.4 Nanomechanical Analysis

The Hysitron TI 950 tribometer, made in the USA, fitted with a Berkovich diamond indenter having a tip radius of 100 nm, was used to measure the nanomechanical properties through nanoindentation and nanoscratch tests. Following Oliver and Pharr's methods, the nanoindentation test sought to determine the hardness (*Hi*) and elastic modulus (*Ei*) of both the substrate and the coating. In these experiments, a load control method was used, where the load was progressively increased at a rate of 600 N/s up to a peak load of 6,000 N within 10 seconds, held there for 2 seconds, and then gradually decreased (at 600 N/s) to zero over an additional 10 seconds.

On the other hand, the nanoscratch test was carried out using the same indenter under a constant load of 8,000 N over a 10 m scratch length (*l*) at a scratching speed of 0.76 m/s. Scanning probe microscopy (SPM) was used to analyze the wear tracks left by the scratches, and Image Metrology A/S, Denmark, provided the SPM image-processing package (SPIPTM) for image processing. The friction coefficient was calculated by dividing the instantaneous lateral force by the normal force. The equations guiding the nanoindentation and nanoscratch tests are described in the literature [29].

9.3 RESULT AND DISCUSSION

9.3.1 Structural Makeup and Morphology of Thin Films

Figure 9.1 shows the TiN film's XRD pattern, which exhibits a single-phase, face-centered cubic (FCC) symmetry. The (111) plane is preferred as its orientation, and this structure displays prominent crystal planes, especially the (111) and (200) planes. Similarly for the Cr-(CrN/TiN) film, the face-centered cubic structure (FCC) was validated by the diffraction pattern analysis [30]. It is important to note that some substrate peaks are evident because the coating is less than 2.5 μm in thickness. Additionally, peaks corresponding to (111), (220), and (311) can be seen in the crystalline structure of the Cr-(CrN/TiN) multilayer film. In the Cr-(CrN/TiN) multilayer film made with the CAE approach, researchers have previously verified the coexistence of TiN and CrN phases [31].

Numerous macro-particles can be seen on both the TiN and CrN/TiN coatings in the SEM images of the coated surfaces in Figure 9.2(a, b). Similar macro-particles have also been seen in coatings made of TiN [32] and TiN/CrN [33] that were applied via CAE-PVD. Although these macro-particles are nearly the same size in both coatings, the TiN single-layer coating has more of them than the CrN/TiN multilayer. In contrast to a multilayer coating, the single-layer coating's surface roughness is increased by the presence of macro-particles.

AFM scanning pictures of the surfaces of the TiN single-layer and CrN/TiN multilayer specimens are shown in Figure 9.3. It is obvious that the CrN/TiN coating's surface is a little bit smoother than that of the TiN coating. Previous studies [34] have discovered that CrN coatings grow in a coaxial pattern and have fewer macroparticles than TiN layers, both of which contribute to improved surface smoothness. A qualitative comparison between single-layer and multilayer coatings is also

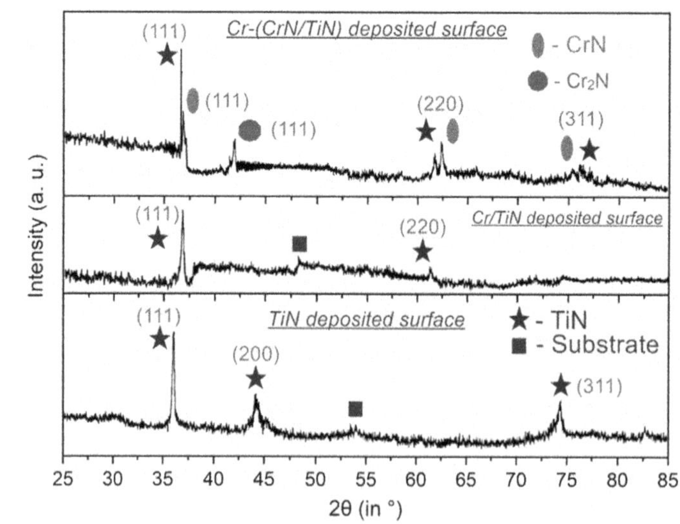

FIGURE 9.1 XRD pattern for (a) TiN, (b) Cr/TiN, and (c) Cr-(Crn/TiN) films.

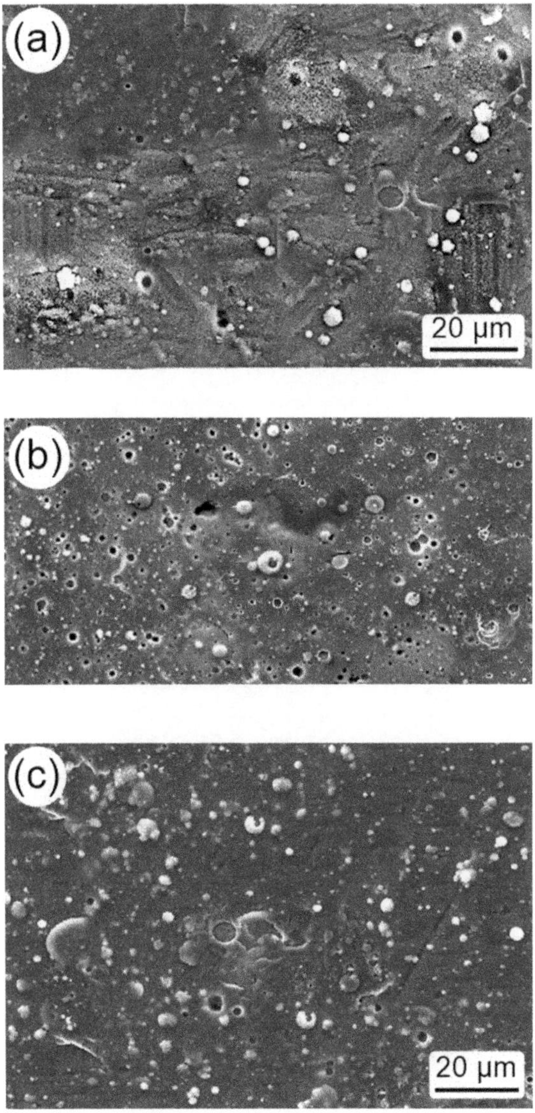

FIGURE 9.2 SEM morphological image of (a) TiN, (b) Cr/TiN, and (c) Cr-(CrN/TiN) films.

provided by the two three-dimensional photographs in Figure 9.3, which unmistakably reveal that CrN/TiN multilayer coatings display superior surface smoothness than TiN single-layer coatings.

The surface roughness of coatings made of TiN, Cr/TiN, and Cr-(CrN/TiN) is compared in Table 9.1 between the three materials. Ra stands for the roughness value, Rz for the average of the longest roughness values from five measurements, and Rmax for the height of the highest roughness. The numerical values for the Cr-(CrN/TiN) are lower than those for the TiN and Cr/TiN films for all three criteria, Ra, Rz,

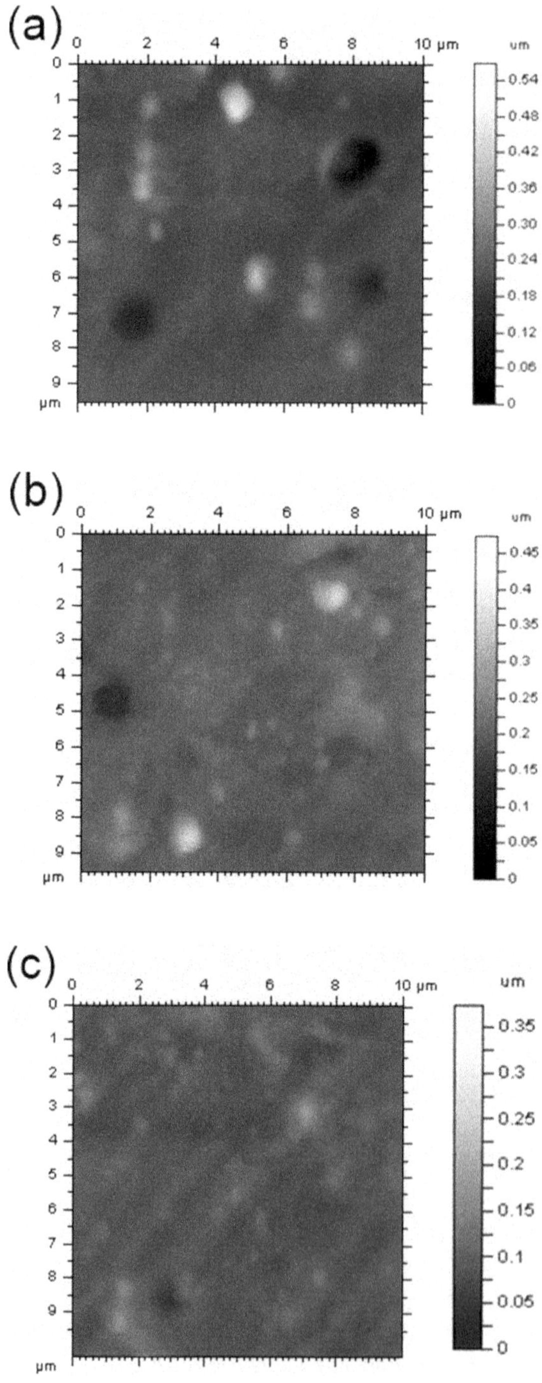

FIGURE 9.3 AFM images of (a) TiN, (b) Cr/TiN, and (c) Cr-(CrN/TiN) films.

TABLE 9.1
Surface roughness parameters and their values for TiN, Cr/TiN, and Cr-(CrN/TiN) films

	Height Parameters	TiN	Cr/TiN	Cr-(CrN/TiN)
Sq	Root mean square height (in μm)	0.0317	0.0247	0.0189
Sa	Arithmetic mean height (in μm)	0.0182	0.0167	0.0124
Sp	Maximum peak height (in μm)	0.261	0.243	0.183
Sz	Maximum height (in μm)	0.537	0.492	0.349
Sv	Minimum pit height (in μm)	0.279	0.251	0.198
Sku	Kurtosis	16.4	13.7	10.4
Ssk	Skewness	1.61	1.24	0.87

and Rmax, indicating a larger region of surface smoothness for the Cr-(CrN/TiN) multilayer film. The roughness values are smaller for Cr/TiN film compared to TiN without interlayer, and a similar result has also been reported in the literature [35].

9.3.2 Nanomechanical Properties of Thin Films

i) Nanoindentation

The maximum penetration depths (h_{max}) for the TiN, Cr/TiN, and Cr-(Cr/TiN) films are 112±4 nm, 104±3 nm, and 86±6nm, respectively, according to the load-displacement (L-D) curve shown in Figure 9.4. According to Equation (1), both the contact area (A) and contact depth (h_c) affect the substrate's hardness. Equation (2) describes the link between the substrate's elastic modulus (E_c) and the reduced elastic modulus (E_{rc}). Table 9.2 presents comprehensive data on the residual depth (h_r) and nanomechanical characteristics of the TiN, Cr/TiN, and Cr-(Cr/TiN) films.

$$H_i = P_{max} / A(h_c) \tag{1}$$

Where P_{max} is the maximum load applied during indentation, A is the contact area, and h_c is the contact depth.

$$\frac{1}{E_r} = \frac{(1-v^2)}{E} + \frac{(1-v_i^2)}{E_i} \tag{2}$$

Here, E_i and v_i stand for the elastic modulus and Poisson's ratio of the sample, respectively, whereas E_{in} stands for the indenter's elastic modulus (1141 GPa), and v_{in} stands for the indenter's Poisson's ratio (0.07).

The nano hardness of the TiN film increased by about 45.12%, whereas the elastic modulus increased by 17.58%, according to the results of the nanoindentation test summarized in Table 9.2. Minimal residual depths, both Cr/TiN and Cr-(CrN/TiN)

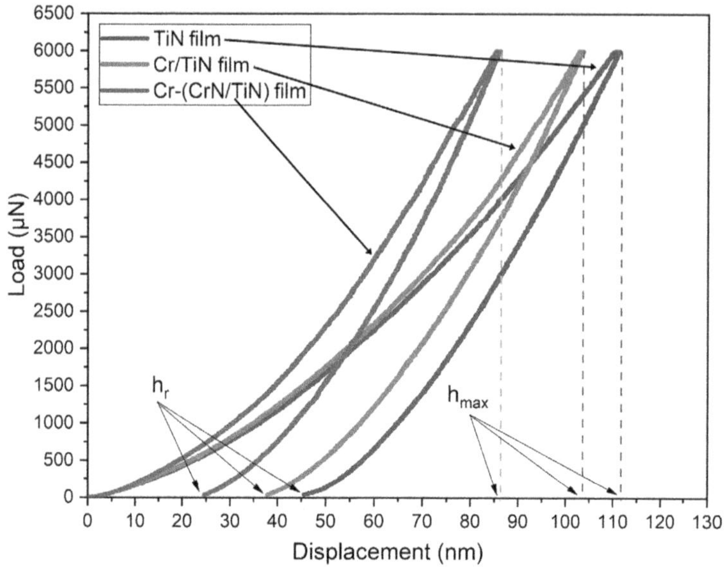

FIGURE 9.4 Displacement curve with respect to an applied load.

TABLE 9.2
Nanomechanical properties of TiN films with and without interlayer

Substrate	Residual Depth, hr (in nm)	Maximum Penetration Depth, h_{max} (in max)	Indentation Hardness, Hi (in GPa)	Indentation Elastic Modulus, Ei (in GPa)	Wear Resistance, Hi/Ei	Plastic Deformation Resistance, H_i^3 / E_i^2 (in GPa)
TiN	45±3	112±4	21.5±1.2	273±12	0.079±0.001	0.133±0.011
Cr/TiN	37±4	104±3	25.7±1.8	298±10	0.086±0.004	0.191±0.026
Cr-(CrN/TiN)	24±5	86±6	31.2±1.6	321±8	0.097±0.002	0.295±0.03

showed considerably less plastic deformation. Additionally, after Cr and Cr-CrN deposition, the TiN film's fracture toughness properties (H/E and H^3/E^2) significantly improved (Table 9.3). The L-D curve's structure shows that Cr/TiN and Cr-(CrN/TiN) films display inelastic behavior, even showing pop-ins at greater indentation loads, perhaps as a result of grain boundary effects in the indenter's stress field. Dislocation nucleation in a neighboring grain and slip transfer may be the predominant mechanisms [36].

Huang et al. [37] used Ti as an interlayer between the substrate and TiN film and discovered that the film thickness had little effect on the nano hardness of the

TABLE 9.3
Outcomes of nanoscratch test for different films

Substrate	Scratch Width (in, μm)	Scratch Hardness (in, *Hs*)	Wear Rate (in, m^2N^{-1})	Average CoF
TiN	3.7±0.6	26.7±1.4	$1.78\times10^{-12} \pm 0.2\times10^{-12}$	0.53±0.05
Cr/TiN	1.8±0.3	31.4±1.15	$4.47\times10^{-13} \pm 0.7\times10^{-13}$	0.41±0.02
Cr-(CrN/TiN)	1.1±0.1	37.3±0.8	$2.31\times10^{-13} \pm 0.41\times10^{-13}$	0.34±0.03

film. Additionally, they noted that the samples' residual stress dropped as TiN film thickness was increased. Utilizing a related interlayer, Yuan et al. [38] found that the indentation stress of TiN with a Ti interlayer was primarily localized around the indenter tip, with the top TiN layer experiencing the highest stress. The stress in the coating decreased as the number of layers grew, while plastic deformation increased. The multilayer coating with the highest thickness ratio exhibited the smallest plastic deformation but generated the most stress during indentation. Similar results were reported by Sheng et al. [39], who found that the elastic modulus and hardness of the TiN multilayer film increased initially and subsequently dropped as the number of layers grew.

ii) Nanoscratch

Figure 9.5 displays images taken using a scanning probe microscope of nanoscratch testing for TiN-, Cr/TiN-, and Cr-(CrN/TiN)-deposited films. These photographs demonstrate that the indenter produced a ploughing effect while scratching as it moved. As a result, it dug up material, which later piled up at the other end to form a pile. The groove's form mimics that of the indenter (see Figure 9.5).

Figure 9.5(c) for Cr-(CrN/TiN) film shows that the groove's width and depth gradually decreased from sample TiN to sample Cr/TiN film. In the instance of Cr-(CrN/TiN) film, pile-up formation is essentially nonexistent. The TiN film with Cr interlayer had the smallest scratch-groove width compared to TiN film without interlayer.

In this study, the amount of wear track (groove volume) per unit load per unit sliding distance (m^2N^{-1}) is used to express the wear rate of the thin film deposited materials. Using Equation (3), the wear volume brought on by the scratching process was calculated.

$$V_{wr} = 0.5\cos(70.3)h_{wt}^2 l_{wt} \tag{3}$$

The wear volume (V_{wr}) is computed mathematically as the sum of the depth (h_{wt}) and length (l_{wt}) of the wear track. As a result, Equation (4) can be used to mathematically express the wear rate (WR):

$$WR = \frac{V_{wr}}{LS_d} \tag{4}$$

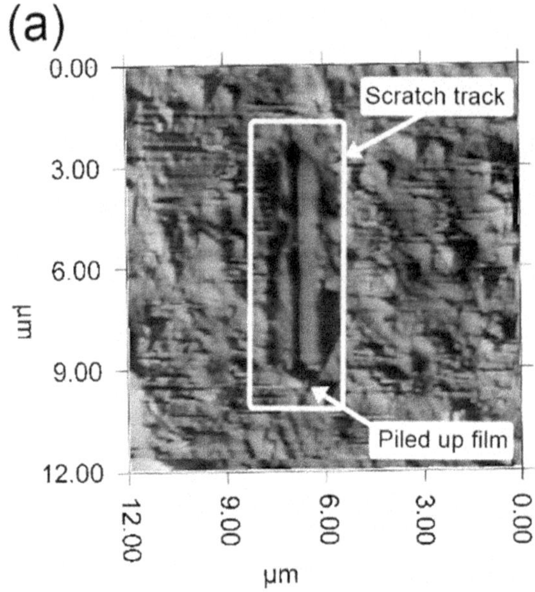

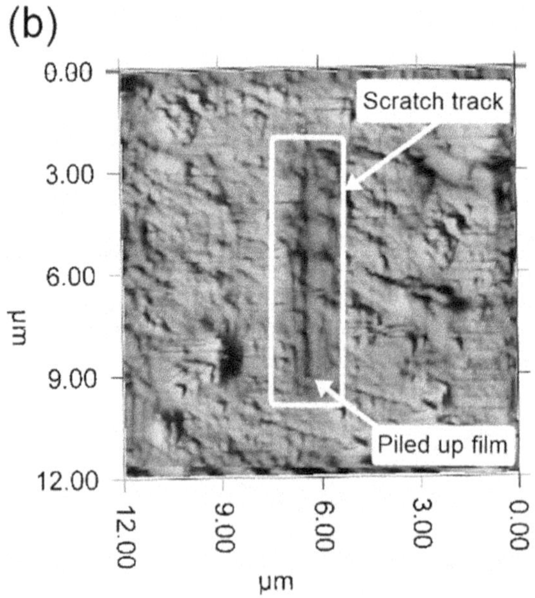

FIGURE 9.5 Scratch track of (a) TiN, (b) Cr/TiN, and (c) Cr-(CrN/TiN) films.

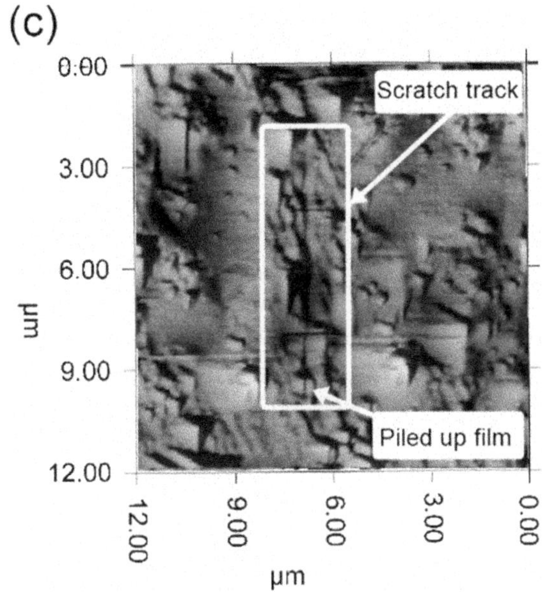

FIGURE 9.5 (Continued)

In this instance, *WR* denotes the wear rate (m^2N^{-1}), *L* is the load (N), and *Sd* is the total sliding distance (m).

The nanoscratch results are presented in Table 9.3, from which it can be noted that TiN film with Cr/CN interlayer has the least wear rate and high scratch hardness compared to TiN film with Cr interlayer and without interlayer. From this observation, it can be said that adhesion between the substrate and TiN film improved as the Cr and Cr/CrN interlayer was introduced. Li et al. [40] have also experienced an improvement in the adhesion strength of TiN film with a Cr interlayer when they compared TiN without and with an interlayer. In other studies, the researchers [41, 42] have noticed that adhesion between the substrate and thin film can be improved using the Cr/CrN film.

The fluctuation in the friction coefficient (CoF) seen in Cr/TiN and Cr-(CrN/TiN) films follows the same logic. The CoF values (Table 9.3) were established using experimental data gathered throughout the scratching procedure. When it comes to Cr-interlayered TiN film, the sample shows a roughly 22.64% reduction in CoF, while Cr-(CrN/TiN) film shows a significant reduction of roughly 35.85% when compared to TiN film without interlayer. It is evident that the use of Cr and CrN films as interlayers, which are still active research topics, has a major impact on the microstructure and characteristics of working layer films that contain hard particles [26, 31, 33, 35].

9.4 CONCLUSIONS

TiN, Cr/TiN, and Cr-(CrN/TiN) films were successfully deposited on hardened DAC-10 tool steel using the CAE-PVD deposition method. Structural, morphological, and adhesion properties of these films were evaluated and compared. From the comparative results, the following conclusions have been noted:

- TiN film with (111) plane orientation was predominant over other peaks (200), (220), and (311), respectively, for all three deposited films. Small peaks of substrate were also detected, and it might be due to the thin film of TiN without any interlayer.
- A huge amount of macroparticles and needle-like cavities were observed on TiN film without any interlayer. The amount of these defects was decreased by using the interlayer (Cr, CrN) between the substrate and TiN film. These defects sometimes are also responsible for deteriorating the surface quality of the final deposited surface. This was validated by the AFM results, which show 8.25% and 31.85% reduction in surface roughness (Ra) value of TiN film after using Cr and Cr/CrN interlayer respectively.
- Nanomechanical properties were also improved after using the Cr and Cr/CrN interlayer between substrate and TiN film. Residual depth decreases by 17.78% and 46.67%, while maximum penetration depth decreases by 7.14% and 23.21% respectively. This reduction in residual depth and maximum penetration depth improves hardness by 19.53% and 45.12%, whereas Young's modulus improved by 9.16% and 17.58% respectively. The wear resistance and resistance to plastic deformation also improved by 8.86%, 22.79% and 4.36%, 121.8% respectively.
- Cr and Cr/CrN interlayers also improve the nanoscratch properties of TiN film. Scratch width reduced by 51.35% and 70.27%, while wear rate was reduced by 74.89% and 87.02% respectively. Similarly, the scratch hardness was improved by 17.6% and 39.7%, whereas friction coefficient was reduced by 22.64% and 35.85% respectively.

REFERENCES

1. Chandrasekaran M. Forging of metals and alloys for biomedical applications. *Metals for Biomedical Devices*, 2019; 293–310.
2. Emamverdian AA, Sun Y, Cao C, Pruncu C, Wang Y. Current failure mechanisms and treatment methods of hot forging tools (dies)-a review. *Engineering Failure Analysis*, 2021; 129: 105678.
3. Nurbanasari M, Tsakiropoulos P, Palmiere EJ. Microstructural evolution of a heat-treated H23 tool steel. *International Journal of Minerals, Metallurgy, and Materials*, 2015; 22: 272–284.
4. Podgornik B, Paulin I, Zajec B, Jacobson S, Leskovšek V. Deep cryogenic treatment of tool steels. *Journal of Materials Processing Technology*, 2016; 229: 398–406.
5. Díaz-Guillén JC, Naeem M, Hdz-García HM, Acevedo-Davila JL, Díaz-Guillén MR, Khan MA,. . . Mtz-Enriquez AI. Duplex plasma treatment of AISI D2 tool steel by combining plasma nitriding (with and without white layer) and post-oxidation. *Surface and Coatings Technology*, 2020; 385; 125420.
6. Chang YY, Amrutwar S. Effect of plasma nitriding pretreatment on the mechanical properties of AlCrSiN-coated tool steels. *Materials*, 2019; 12(5): 795.
7. Miyamoto J, Abraha P. The effect of plasma nitriding treatment time on the tribological properties of the AISI H13 tool steel. *Surface and Coatings Technology*, 2019; 375: 15–21.

8. Yazıcı M, Kovacı H, Yetim AF, Çelik A. Structural, mechanical and tribological properties of Ti and TiN coatings on 316L stainless steel. *Ceramics International*, 2018; 44(12): 14195–14201.
9. Du JW, Chen L, Chen J, Du Y. Mechanical properties, thermal stability and oxidation resistance of TiN/CrN multilayer coatings. *Vacuum*, 2020; 179: 109468.
10. Hu X, Dong C, Wang Q, Chen B, Yang H, Wei T,. . . Chen D. High-temperature oxidation of thick Cr coating prepared by arc deposition for accident tolerant fuel claddings. *Journal of Nuclear Materials*, 2019; 519: 145–156.
11. Brachet JC, Idarraga-Trujillo I, Le Flem M, Le Saux M, Vandenberghe V, Urvoy S,. . . Sanchette F. Early studies on Cr-Coated Zircaloy-4 as enhanced accident tolerant nuclear fuel claddings for light water reactors. *Journal of Nuclear Materials*, 2019; 517: 268–285.
12. Xiao W, Deng H, Zou S, Ren Y, Tang D, Lei M,. . . Chen Y. Effect of roughness of substrate and sputtering power on the properties of TiN coatings deposited by magnetron sputtering for ATF. *Journal of Nuclear Materials*, 2018; 509: 542–549.
13. Sivakumar ER, Krishna R, Senthilkumar P, Colak I. Influence of mechanical properties of coating and substrate on wear performance of HDLC and TiN-coated AISI 5140 alloy steel. *Proceedings of the Institution of Mechanical Engineers, Part E: Journal of Process Mechanical Engineering*, 2022; 09544089221132434.
14. Alat E, Motta AT, Comstock RJ, Partezana JM, Wolfe DE. Ceramic coating for corrosion (c3) resistance of nuclear fuel cladding. *Surface and Coatings Technology*, 2015; 281: 133–143.
15. Bharathi SS, Ravindran D, Moshi AAM, Alagarsamy SV, Kannan RK, Prasath VJ. Machinability study on CNC turning of stainless steel 303 with CVD multi-layer (TiN/Al2O3/TiCN) coated carbide insert by using grey-fuzzy logic approach. *Proceedings of the Institution of Mechanical Engineers, Part E: Journal of Process Mechanical Engineering*, 2022; 236(5): 1967–1978.
16. Meng C, Yang L, Wu Y, Tan J, Dang W, He X, Ma X. Study of the oxidation behavior of CrN coating on Zr alloy in air. *Journal of Nuclear Materials*, 2019; 515: 354–369.
17. Park DJ, Kim HG, Jung YI, Park JH, Yang JH, Koo YH. Microstructure and mechanical behavior of Zr substrates coated with FeCrAl and Mo by cold-spraying. *Journal of Nuclear Materials*, 2018; 504: 261–266.
18. Yeom H, Maier B, Johnson G, Dabney T, Walters J, Sridharan K. Development of cold spray process for oxidation-resistant FeCrAl and Mo diffusion barrier coatings on optimized ZIRLO™. *Journal of Nuclear Materials*, 2018; 507: 306–315.
19. Bischoff J, Vauglin C, Delafoy C, Barberis P, Perche D, Guerin B,. . . Brachet JC. Development of Cr-coated zirconium alloy cladding for enhanced accident tolerance. In *Topfuel 2016-Light Water Reactor (LWR) Fuel Performance Meeting*, 2016.
20. Brachet JC, Le Saux M, Lezaud-Chaillioux V, Dumerval M, Houmaire Q, Lomello F,. . . Pouillier E. Behavior under LOCA conditions of enhanced accident tolerant chromium coated zircaloy-4 claddings. In *Topfuel 2016-Light Water Reactor (LWR) Fuel Performance Meeting*, 2016.
21. Park JH, Kim HG, Park JY, Jung YI, Park DJ, Koo YH. High temperature steam-oxidation behavior of arc ion plated Cr coatings for accident tolerant fuel claddings. *Surface and Coatings Technology*, 2015; 280: 256–259.
22. Kuprin AS, Belous VA, Voyevodin VN, Bryk VV, Vasilenko RL, Ovcharenko VD,. . . V'yugov PN. Vacuum-arc chromium-based coatings for protection of zirconium alloys from the high-temperature oxidation in air. *Journal of Nuclear Materials*, 2015; 465: 400–406.
23. Su R, Zhang H, O'Connor DJ, Shi L, Meng X, Zhang H. Deposition and characterization of Ti2AlC MAX phase and Ti3AlC thin films by magnetron sputtering. *Materials Letters*, 2016; 179: 194–197.

24. Chen ZS, Li HJ, Fu QG, Yang D, Shen QL. Annealing effect on microstructure and mechanical properties of magnetron sputtering Ti–Si–C thin film. *Materials Science and Technology*, 2013; 29(8): 975–979.
25. Kim HG, Kim IH, Jung YI, Park DJ, Park JY, Koo YH. Adhesion property and high-temperature oxidation behavior of Cr-coated Zircaloy-4 cladding tube prepared by 3D laser coating. *Journal of Nuclear Materials*, 2015; 465: 531–539.
26. Xiao W, Chen H, Liu X, Tang D, Deng H, Zou S,. . . Lei M. Thermal shock resistance of TiN-, Cr-, and TiN/Cr-coated zirconium alloy. *Journal of Nuclear Materials*, 2019; 526: 151777.
27. Kumar S, Maity SR, Patnaik L. Morphology and wear behavior of monolayer TiAlN and composite AlCrN/TiAlN-coated plasma-nitrided DAC-10 tool steel. *Arabian Journal for Science and Engineering*, 2022; 47(12): 15519–15538.
28. Lousa A, Romero J, Martınez E, Esteve J, Montala F, Carreras L. Multilayered chromium/chromium nitride coatings for use in pressure die-casting. *Surface and Coatings technology*, 2001; 146: 268–273.
29. Kumar S, Maity SR, Patnaik L. Friction and tribological behavior of bare nitrided, TiAlN and AlCrN coated MDC-K hot work tool steel. *Ceramics International*, 2020; 46(11), 17280–17294.
30. Wang Y, Northwood DO. Effect of substrate material on the corrosion of TiN-coated stainless steels in simulated anode and cathode environments of proton exchange membrane fuel cells. *Journal of Power Sources*, 2009; 191(2): 483–488.
31. Ou YX, Lin J, Tong S, Sproul WD, Lei MK. Structure, adhesion and corrosion behavior of CrN/TiN superlattice coatings deposited by the combined deep oscillation magnetron sputtering and pulsed dc magnetron sputtering. *Surface and Coatings Technology*, 2016; 293: 21–27.
32. Sanchette F, Ducros C, Schmitt T, Steyer P, Billard A. Nanostructured hard coatings deposited by cathodic arc deposition: From concepts to applications. *Surface and Coatings Technology*, 2011; 205(23–24): 5444–5453.
33. Tian LH, Liu EQ, Fan AL, Qin L, Liu DX, Tang B, Pan JD. Effect of TiN/CrN multilayer coating by cathodic arc deposition on wear and corrosion behaviours of AZ91D magnesium alloy. *Materials Science Forum*, 2009; 610: 870–873.
34. Liu C, Bi Q, Leyland A, Matthews A. An electrochemical impedance spectroscopy study of the corrosion behaviour of PVD coated steels in 0.5 N NaCl aqueous solution: Part II.: EIS interpretation of corrosion behaviour. *Corrosion science*, 2003; 45(6): 1257–1273.
35. Nam ND, Kim MJ, Jo DS, Kim JG, Yoon DH. Corrosion protection of Ti/TiN, Cr/TiN, Ti/CrN, and Cr/CrN multi-coatings in simulated proton exchange membrane fuel cell environment. *Thin Solid Films*, 2013; 545: 380–384.
36. Pöhl F. Pop-in behavior and elastic-to-plastic transition of polycrystalline pure iron during sharp nanoindentation. *Scientific reports*, 2019; 9(1):15350.
37. Huang JH, Ouyang FY, Yu GP. Effect of film thickness and Ti interlayer on the structure and properties of nanocrystalline TiN thin films on AISI D2 steel. *Surface and Coatings Technology*, 2007; 201(16–17): 7043–7053.
38. Yuan Z, Han Y, Zang S, Chen J, He G, Chai Y,. . . Fu Q. Analysis of the mechanical properties of TiN/Ti multilayer coatings using indentation under a broad load range. *Ceramics International*, 2021; 47(8): 10796–10808.
39. Sheng L, Xiao Y, Jiao C, Du B, Li Y, Wu Z, Shao L. Influence of layer number on microstructure, mechanical properties and wear behavior of the TiN/Ti multilayer coatings fabricated by high-power magnetron sputtering deposition. *Journal of Manufacturing Processes*, 2021; 70: 529–542.
40. Li D, Guruvenket S, Hassani S, Bousser E, Azzi M, Szpunar JA, Klemberg-Sapieha JE. Effect of Cr interlayer on the adhesion and corrosion enhancement of nanocomposite

TiN-based coatings deposited on stainless steel 410. *Thin Solid Films*, 2011; 519(10): 3128–3134.
41. Arias DF, Gómez A, Souza RMD, Vélez JM. Residual stress gradient of Cr and CrN thin films. *Materials Chemistry and Physics*, 2018; 204: 269–276.
42. Cai X, Gao Y, Cai F, Zhang L, Zhang S. Effects of multi-layer structure on microstructure, wear and erosion performance of the Cr/CrN films on Ti alloy substrate. *Applied Surface Science*, 2019; 483: 661–669.

10 Structural and Physiomechanical Evaluation of AlCr(Si)N Thin Films Coatings

Lokeswar Patnaik and Magdalena Łępicka

10.1 INTRODUCTION

The challenge of enhancing the wear resistance of metallic implants has driven the adoption of surface-modification strategies, with the design and optimization of anti-wear thin films emerging as a vibrant and enduring field of research. Among others, classic transition nitrides such as CrN have gone through constant alterations and modifications. One such modification is the introduction of metallic elements such as Ti [1], Al [2], Hf [3], and V [4], semi-metallic elements such as Si [5] and B [6], and non-metallic elements such as O [7] and C [8] in CrN matrix to form complex multicomponent films leading to the formation of its ternary and quaternary counterparts. One such modification in CrN with Al addition forms a popular ternary coating namely AlCrN which, compared to CrN film, possesses superior hardness, anti-wear properties, and resistance to oxidation at elevated temperatures up to 700 °C [9, 10]. The mechanical properties of this film are characterized by the Al content in it, which influences the metastable Al-Cr-N solid solution and AlN phase in the form of cubic, or hexagonal, or wurtzite structure. An increase in Al content results in hcp structure of the film, which is characterized by inferior mechanical properties compared to the fcc structure [11]. Further, the properties of AlCrN can be tailored during deposition, among others, the Al-Cr target composition [11], nitrogen pressure [12], and substrate bias voltage [13]. Reiter et al. [11] obtained an increase in hardness to about 30% in AlCrN film with 70% Al content. AlCrN film can be synthesized via various deposition techniques, including, for example, cathodic deposition HIPIMS, ion arc plating, and magnetron sputtering among others.

However, the cathodic deposition technique is opted more often owing to higher degree of plasma ionization and energy resulting in rich density of deposited material compounding to improved mechanical properties and adhesion. Although one of the disadvantages of cathodic deposition is tad inferior surface quality attributed to the formation of micro droplets and craters. Interestingly, deposition made through magnetron sputtering results in better surface quality; however, the plasma density during deposition is low propounding to inferior mechanical properties in the film compared

 DOI: 10.1201/9781003750659-10

to cathodic deposition [14]. Further, another important semi-metallic element is silicon whose introduction in Al-Cr-N matrix forms a quaternary coating system. Its addition forms a solid solution and induces grain hardening phenomenon. This compounds to grain refinement which results in improved hardness in accordance with Hall-Petch effect. Nonetheless, silicon has minimal solubility in Al-Cr-N matrix and an excess amount in the crystal lattice causes segregation in the grain boundary owing to formation of more amorphous Si_xN_y phase, which ultimately results in decrement in the mechanical properties of the film [15]. In this chapter, the morphology including structural makeup and topography of AlCr(Si)N and their mechanical performance are comprehensively studied.

10.2 MATERIALS AND METHODS

The deposition was performed on medical-grade 316 LVM stainless steel substrate (ϕ 30 mm × 5 mm disc) using cathodic arc evaporation technique (Ingenious Tools and Design Solution, India). Prior to deposition, the samples were ground and subjected to a series of polishing (abrasive paper 200–1,000 grit size) followed by lapping using diamond paste to obtain a smooth and polished surface. All the samples (or substrates) were cleaned ultrasonically in alkaline medium, rinsed in de-ionized water, and dried in vacuum. The mean surface roughness, hardness, and elastic modulus of the polished 316 LVM substrate were 0.6±0.11 μm, 3.2±0.8 GPa, and 198±18 GPa respectively. A CrN adhesive interlayer was deposited using Cr cathode (purity 99.99%) prior to the deposition of the functional coatings. The chamber was evacuated to a pressure of 4×10^{-3} Pa, and the substrate was heated and maintained at a temperature of 380 °C. High-purity argon was infused at 30 sccm flow rate to get to a chamber pressure of 2×10^{-2} Pa. Subsequently, nitrogen gas was introduced in the chamber with 800 sccm flow rate at 2 N pressure. The substrate bias voltage and current were kept at −80 V and 90 A respectively. The deposition was carried out at a pressure of 4.1×10^{-2} Pa for 30 minutes to obtain a 1-μm-thick CrN layer. The chrome-nitrided substrate was further polished to obtain a smooth surface over which AlCrN and AlCrSiN topcoats were deposited using $Al_{70}Cr_{30}$ and $Al_{68}Cr_{30}Si_2$ targets. The chamber pressure was controlled at 1×10^{-3} Pa and the substrate was heated to 400 °C. The chamber was etched with Ar for 15 minutes at a substrate voltage of −200 V. For the deposition, the arc current and substrate bias voltage were 100 A and −100 V respectively. Nitrogen was introduced inside the chamber at 4 Pa controlled by capacitive gauges. After 45 minutes, a 1.5-μm-thick AlCrN and AlCrSiN film was obtained on the chrome-nitrided 316 LVM substrate. The cross-sectional and surface feature of the coatings was observed using scanning electron microscopy (VP:55, Zeiss Supra, Germany) coupled with energy dispersive spectrum attachment for elemental analysis. The structural makeup of the coatings was evaluated using X-ray diffraction (XRD) technique (X'Pert³ MRD, Malvern Panalytical, India) using CuKα radiation of 1.54 Å wavelength produced with a voltage and current of 40 kV and 25 mA respectively. The scanning was performed with 0.05° step size on 30° to 70° scanning range (2θ). The surface topography of the film was analyzed using atomic force microscopy (model: 5100, Agilent Technologies, USA) for an area of 20 μm × 20 μm using monolithic silicon tip at a scanning speed of 512 points/s.

The phase and polymorphic analysis of the coatings was conducted using Raman spectroscopy integrated with He-Ne laser (λ= 632.8 nm) focused with a numerical aperture of 0.5 (HR800, Horiba Jobin Vyon, Japan). The spectrum was evaluated via 600 lines/mm grating. The light detection was done using a silicon-charged camera. Nanoindentation test was conducted on the coatings to evaluate the mechanical attributes using Berkovich indenter of 100 nm diameter at a loading rate of 20 μN/s to a peak load of 1,000 μN (TI950 Triboindenter, Hysitron, Bruker, India). Upon loading, the indenter was held for 5 s to access the creep nature of the coating material post which it was unloaded to obtain the P-h curve. The scratch behavior of the coatings was accessed through nanoscratch test using the same indenter using constant load of 2,000 μN at a scratching speed of 0.76 μm/s. The topography of the indentation and scratch on the coatings was analyzed by scanning probe microscopy (SPM) using SPIP™ image processing software (A/S, Image metrology, Denmark). Ten number of experiments were repeated in each category to obtain a reliable statistic. The governing equations for obtaining the particulars of the nanoindentation and nanoscratch test following Oliver and Pharr method is categorically presented in a study by Sribalaji et al. [16]. The adhesion of the films was evaluated by Rockwell-C indentation test (model: RASNET-3, MTMS, India) using a diamond conical tip of 0.2 mm radius and 120° tip radius following VDI 3198 guidelines. The indentation was performed at 1,500 N load, and the impression on the films was analyzed using scanning electron microscope to classify the quality ranking in accordance with Daimler-Benz adhesion quality rankings.

10.3 RESULTS AND DISCUSSION

10.3.1 Morphology of the Films

The cross-sectional SEM micrograph of AlCrN and AlCrSiN with the corresponding line EDS spectrum is presented in Figure 10.1(a-a′) and (b-b′) respectively. The microstructure of the both the coatings appeared dense with a distinct CrN interface. However, AlCrN film showcased a typical columnar structure which got denser with the introduction of Si in the Al-Cr-N matrix attributed to nanocrystalline growth of Al and Si in the film. The formation of such nanocomposite structure is piloted by the limiting of total system free energy by virtue of large cohesive energy at the interface despite the formation of amorphous Si_xN_y phases in AlCrSiN film. Lee and Chang [17] observed that Si-addition reduces the formation of columnar structure, and the resulting film comprises nanocrystalline CrN surrounded by amorphous Si_xN_y phases. As a matter of fact, the formation of complex nanocomposite ternary or quaternary structure is favored by higher mixing enthalpy [18]. Also, both the coatings did not show any evidence of internal fracture or cracks indicating homogeneity in the compound formation during deposition. Further, Si-addition propounds to grain growth contributing to recrystallization-induced equi-axial grain growth, which subsequently retards the columnar structure of AlCrSiN. A similar observation in a study by He et al. [19] also showcased a columnar structure in AlCrN and featureless and dense morphology in AlCrSiN film. Previous studies [20, 21] have reported reduction of columnar structure leading to improved resistance to oxidation in Si-incorporated

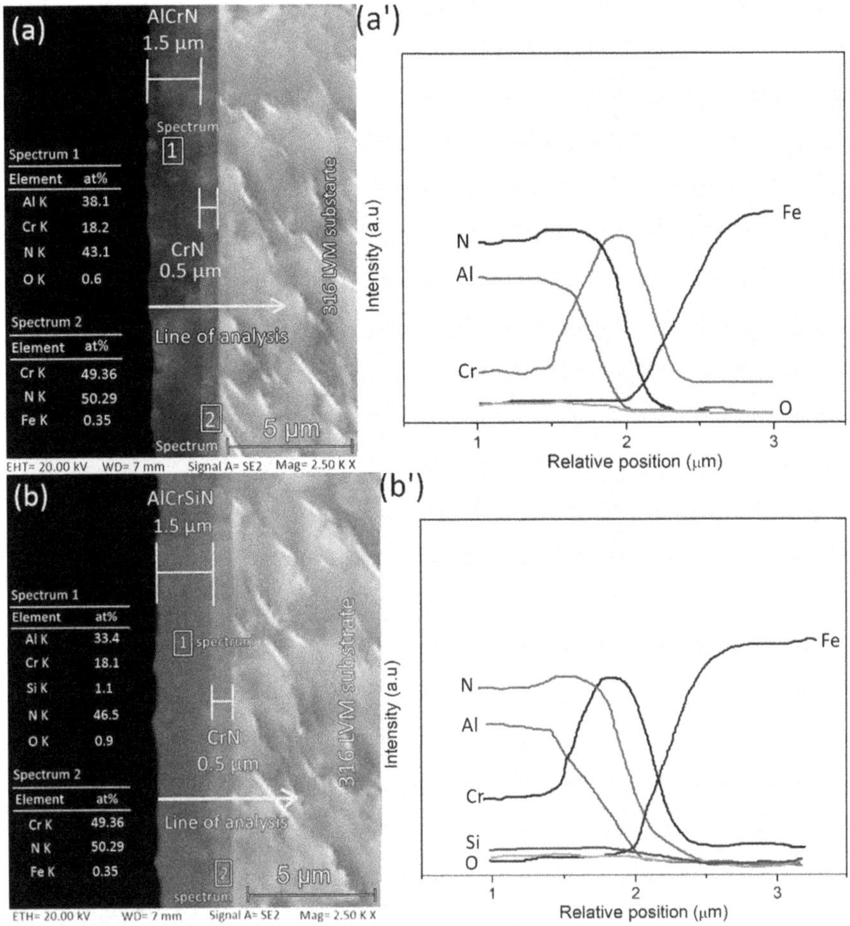

FIGURE 10.1 Cross-sectional SEM image and corresponding line EDS spectrum of (a-a') AlCrN and (b-b') AlCrSiN film.

AlCrN film. The film thickness was measured at ten different points across the cross section and found to be 1.5±0.3 μm and 1.5±0.4 μm for AlCrN and AlCrSiN topcoat respectively. Here, from the point EDS data (see Figure 10.1(a) and (b)), it is interesting to note that the atomic concentration of Al, Cr, and Si in the coatings was lower than the targets ($Al_{70}Cr_{30}$ and $Al_{68}Cr_{30}Si_2$), attributed to evaporation efficiencies of the elements. Liu et al. [22] observed a similar phenomenon with evaporation efficiencies of Al, Cr, and Si in the order of Cr > Al > Si. In another research, Bogdan et al. [23] had suggested that lower atomic concentration in the Al-Cr-N-based coatings (than target) is ascribed to lower atomic mass of Al resulting in more dispersion during collision with nitrogen resulting in lower density in the vapor during deposition. In the present study, the Al/Al+Cr ratio from target to coating was from 0.7 to 0.675 in AlCrN film and 0.694 to 0.652 in AlCrSiN film.

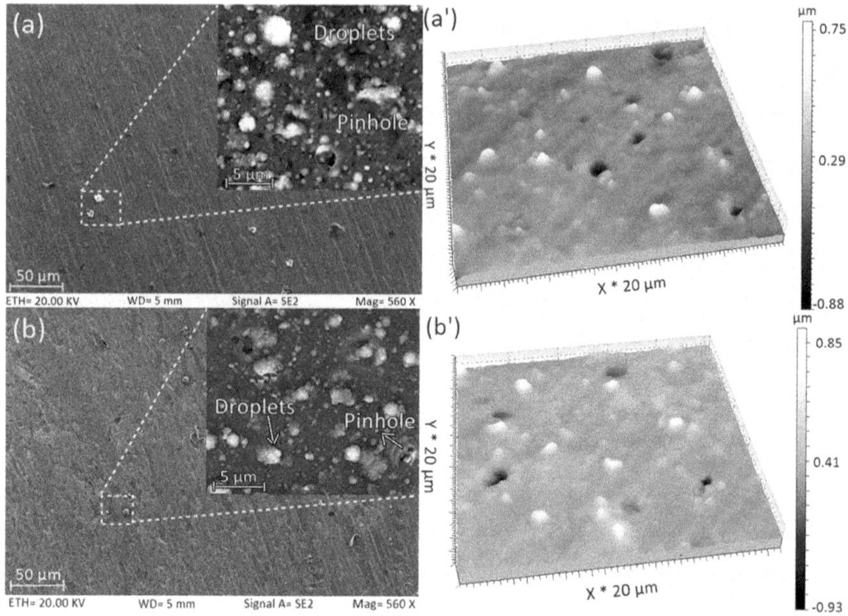

FIGURE 10.2 Surface topography using SEM and AFM micrographs of (a-a′) AlCrN and (b-b′) AlCrSiN.

The SEM topography and AFM micrographs of the AlCrN and AlCrSiN surface are presented in Figure 10.2(a-a′) and (b-b′) respectively. Both coatings exhibited droplets measuring 1–4 μm in diameter, originating from the ejection and coalescence of molten particles at high-temperature cathode arc spots onto the substrate – a characteristic defect of cathodic arc deposition. The average surface roughness (Sa) obtained from the AFM micrographs of AlCrN (see Figure 10.2(a′)) and AlCrSiN (see Figure 10.2(b) was 0.29±0.11 μm and 0.41±0.16 μm respectively. Owing to the formation of micro-droplets, the surface roughness measurement was burdened by high standard deviation values. The surface quality of AlCrSiN appeared slightly inferior than AlCrN; however, there was no statistically significant difference ($p<0.05$) between the average surface roughness of the films. On the contrary, the surface roughness of chrome nitrided surface was 0.21±0.09 μm, which worsened post AlCrN and AlCrSiN deposition. A similar outcome with regard to ternary and quaternary coatings was observed by Bogdan et al. [23].

The diffraction pattern of AlCrN and AlCrSiN presented in Figure 10.3(a) shows B1 cubic structure CrN (c-CrN) in accordance with ICDD 04-004-6868 pattern. Here, it is important to note that no AlN phases (w/c/h-AlN) was detected propounding to total solubility of Al-Cr-N in the NaCl-type microstructure [23]. However, with AlCrSiN, the diffraction lines were slightly widened, indicating a decrease in crystallite size and lattice expansion. No peaks associated with Si_3N_4 were detected in AlCrSiN spectrum, suggesting the presence of Si_3N_4 in amorphous form [24]. It has been reported that amorphous Si_3N_4 imparts improved yield stress and hardness

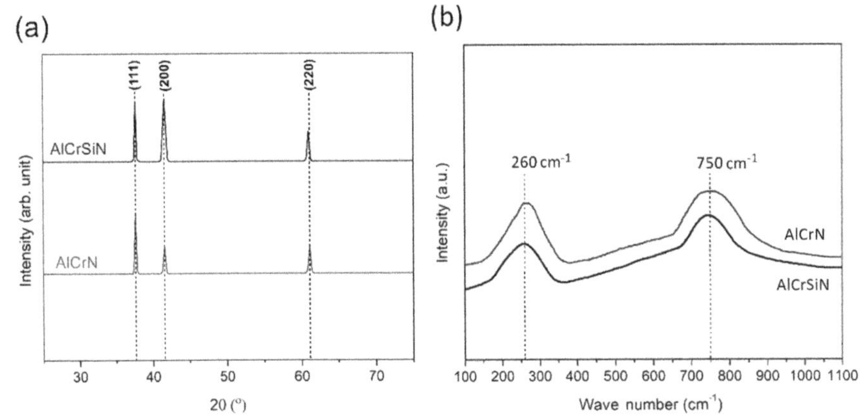

FIGURE 10.3 (a) XRD and (b) Raman spectrum of AlCrN and AlCrSiN films.

to the Si-incorporated AlCrN film [25]. The atomic radius of Si and Al is 0.110 nm and 0.125 nm respectively, it is therefore inferred that introduction of Si may been executed by virtue of replacing the Al sites. The XRD analysis was further corroborated with the Raman bands of the films shown in Figure 10.3(b). Both the films have showcased bands positioned at 260 cm^{-1} and 750 cm^{-1} typically exhibiting an fcc structure. Moreover, there were no characteristics bands at 180 cm^{-1} and 570 cm^{-1} or double band at 190–260 cm^{-1} suggesting the coating was free from wurtzite or fcc + wurtzite structure in the films pointing toward a more stabilized fcc structural makeup [26]. Kaindl et al. [27] reported shifting of Raman bands in Al-Cr-N-based films deposited with higher bias voltage (–100 V to –150 V) is attributed to increased compressive stress in the film. Soldán et al. [26] reported Si addition in expense of Cr sites leads to a higher Al/Cr ratio in the coating, resulting in the destabilization of the fcc structure. They also suggested that increase in bias voltage (> –100 V) results in the stabilization of the fcc structure. In the present work, the bias voltage was kept at – 100 V during deposition. Nonetheless, on account of lower lattice parameter, the crystallite size for CrN (111), CrN (200), and CrN (220) in AlCrN were 17 nm, 7 nm, and 10 nm respectively, which got reduced to 13 nm, 4 nm, and 6 nm in AlCrSiN film.

10.3.2 Mechanical Properties

The p-h plot with corresponding insert of the indentation dent on the surface of AlCrN and AlCrSiN films is presented in Figure 10.4(a) and (b) respectively. Both the films have showcased elastoplastic nature with no evidence of creeping. The results and other particulars of nanoindentation test are presented in Table 10.1. The AlCrSiN film had registered lower residual depth (h_r = 11 nm) and displacement (h_{max} = 38.9 nm) of the indenter propounding to improved hardness and elastic modulus of the film. To put things on perspective, the hardness of AlCrN film was increased by ~26% with Si (at 1.1%) addition. Si incorporation affects the hardness

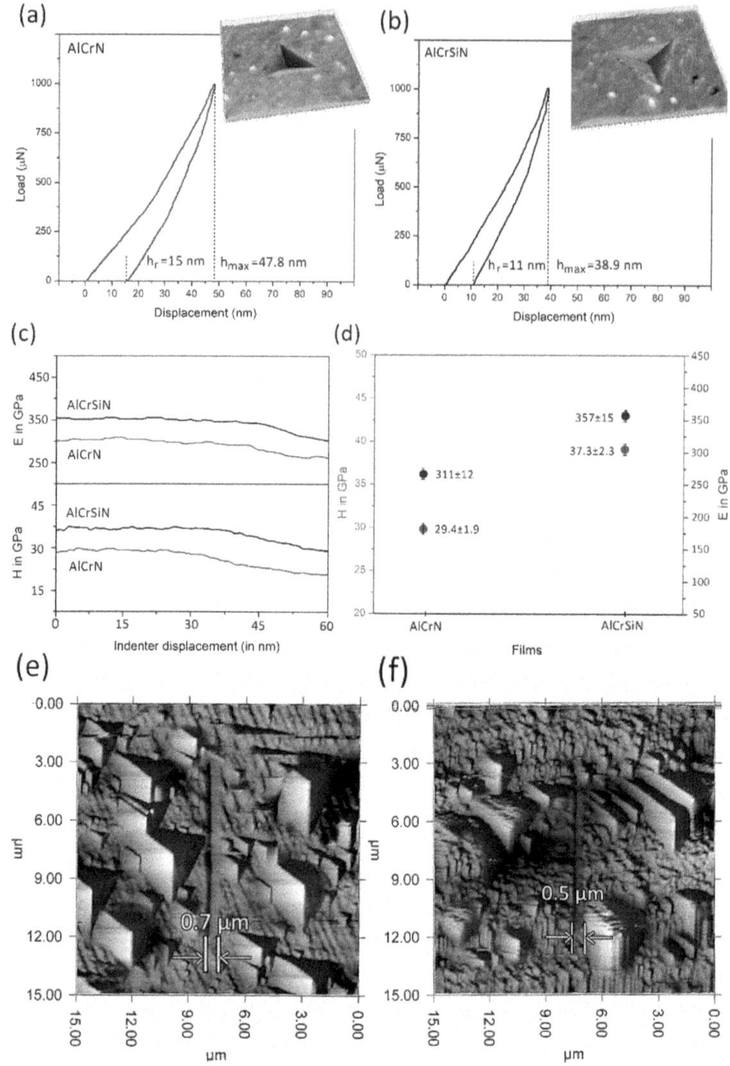

FIGURE 10.4 P-h plot with indentation impression (insert) for (a) AlCrN and (b) AlCrSiN; (c) Trend of H and E along indenter displacement in the films and corresponding (d) error plot; nanoscratch wear track on (e) AlCrN and (f) AlCrSiN films.

and residual stress of the film. One of the plausible reasons to this could be augmented defect generation and grain size refinement inducing hindrance in the grain boundary sliding due to the presence of amorphous Si_3N_4 phase, which insulates at the grain boundary of AlCrN crystal. This phenomenon is corroborated with the findings of S. Veprek and co-workers [28]. However, one of the limitations of the hindering grain boundary discussion is that the dissolution of the Si in the Al-Cr-N nanocrystal structure is not considered. To ensure that dissolution of Si was uniform

TABLE 10.1
Nanoindentation and nanoscratch test results

Tests	**Particulars**	**Thin Films**		**p-Value**	**Tukey's Grouping**
		AlCrN	**AlCrSiN**		
Nanoindentation	Hardness (H in GPa)	29.4±1.9	37.3±2.3	< 0.5	D
	Elastic modulus (E in GPa)	311±12	357±15	< 0.5	D
	Resistance to wear (H/E)	0.094±0.003	0.104±0.004	< 0.5	D
	Resistance to deformation (H^3/E^2 in GPa)	0.26±0.09	0.40±0.11	>0.5	S
	Elastic work (E.W. in pJ)	34±3	46±4	< 0.5	D
	Plastic work (P.W. in pJ)	22±2	18±3	> 0.5	S
	Compressive stress (σ in GPa)	−3.3±0.2	−4.2±0.3	< 0.5	D
Nanoscratch	Scratch hardness (H_s in GPa)	32.3±2.2	43.1±4.5	< 0.5	D
	Friction coefficient	0.19±0.04	0.23±0.06	> 0.5	S

D: Different Tukey's group indicating statistical difference S: Same Tukey's group indicating no statistical difference

in the A-Cr-N matrix, five readings were taken at different places on the coated surface for nanomechanical characterization to ensure data repeatability. Further, the trend of hardness and elastic modulus across the indenter displacement in the films presented in Figure 10.4(c) shows no sudden and abrupt drop in the hardness and elastic modulus indicating homogeneity in the films. The error plot for hardness and elastic modulus response from nanoindentation test is presented in Figure 10.4(d). In addition, increase in residual stress has reportedly improved the hardness response of Al- and Ti-based ternary coatings [29]. Literature suggests that higher bias voltage is a governing parameter for higher residual compressive stress which imparts higher energy to atoms to impinge the surface of the films [30]. From Table 10.1, it can be seen that the compressive residual stress was increased from −3.3 GPa to −4.2 GPa with Si-addition in Al-Cr-N matrix. The compressive residual stress-induced hardness is an outcome of substrate bias voltage during deposition. Ion mixing in coating system instigates stress-fields due to defect generation that acts as a hindrance for dislocation movement leading to grain size refinement in accordance with Hall-Petch effect propounding to improved hardness behavior [23]. Soldán et al. [26] found this phenomenon more pronounced with AlCrSiN films having Si at%≤2 due to better crystallinity. They had reported increased crystallinity and improved hardness in AlCrSiN (2 at.% Si) films deposited with −100 V of bias voltage. Silicon has lower solubility in Al-Cr-N matrix, therefore, its higher content aids in more amorphous Si_xN_y phase formation in the crystal lattice resulting in severe segregation along the grain boundaries compounding to inferior mechanical properties [15, 23, 31]. They had reported that this augmentation ceases at −100 V as it reaches the equilibrium defect density. Also, they had observed that in AlCrSiN having Si at% ≥ 2, the defect saturation ceases at ~−70 V, post which the structure starts to transform into wurtzite phase from fcc structure [26]. Interestingly, the hardness response attributed to Si-addition is dominated by the effect of phase transformation. Many authors [11, 32, 33] have reported drop in the hardness associated with fcc to wurtzite structural transformation. In another work, Schiøtz et al. [34] had pressed the importance of fcc/wurtzite phase ratio on the hardness response rather than Hall-Petch effect-induced reduced grain size and grain boundary hindrance due to Si_3N_4 formation. They concluded that higher fcc/wurtzite ratio (no or lower wurtzite) in the crystalline structure facilitates improved mechanical properties. In the present work, from the Raman spectra of the films (see Figure 10.3(b)), it can be seen that there were no bands (at 190 cm^{-1} and 570 cm^{-1}) associated with wurtzite structure. A slight increase in the elastic modulus was also noted with Si-addition in AlCrN film. Elastic modulus demonstrates the chemical composition in line with the inter-atomic distance between the metastable phases in the material. In short, with increase in elastic modulus, the inter-atomic distance decreases. It is evident with Si-addition in AlCrN, as it was previously seen that the lattice parameter got decreased from 0.4140 nm (AlCrN) to 0.4123 nm (AlCrSiN).

Further, higher H/E is suggestive of improved resistance to wear, and H^3/E^2 ratio demonstrates the onset of plastic deformation in the film, a higher value indicates improved resistance to it. From the results in Table 10.1, higher resistance to wear and plastic deformation is evident with AlCrSiN film. Similar observation in hardness characteristics with Si addition in Al-Cr-N matrix was recorded by Ding et al. [35]. Interestingly, the mechanical behavior of these films is similar across other deposition techniques such as magnetron sputtering [21, 36].

The scratch hardness of the films was obtained from nanoscratch technique. Scratch hardness is an indicator of the adhesion strength of thin films as well [37]. The wear track on AlCrN and AlCrSiN film obtained from SPM is presented in Figure 10.4(e) and (f) respectively. From Table 10.1, it can be seen that the scratch hardness of AlCrSiN (43.1±4.5 GPa) was higher than that of AlCrN (32.3±2.2 GPa); however, the standard deviation registered with the former was higher owing to higher surface roughness value. The width of the wear track was lower in AlCrSiN (0.5 μm) than AlCrN (0.7 μm), attributed to the higher H^3/E^2 factor in the former, inducing better resistance to deform. Interestingly, AlCrSiN film witnessed higher friction coefficient (0.23±0.06) possibly due to higher H/E (0.104±0.004) factor making the film unable to buckle under indenter sliding load due its stiffness and brittleness. Chen et al. [38] reported higher hardness and improved resistance to wear with AlCrSiN film compared to AlTiSiN, attributed to stabilized fcc structure and a higher H^3/E^2 ratio. Further, incorporation of CrN interlayer aids in improved mechanical and anti-scratch properties due to its ability to diffuse with the AlCrN and AlCrSiN topcoat. A similar observation with CrN interlayer for different topcoats is reported in literature [39].

The adhesion of AlCrN and AlCrSiN films was evaluated using Rockwell-C indentation test. The SEM micrographs of the indentation impression on the films is presented in Figure 10.5 (a–b). AlCrSiN film (see Figure 10.5(a)) witnessed the formation of radial cracks which can be categorized with HF1 or HF2 quality rankings. On the other hand, AlCrSiN film (see Figure 10.5(b)) had suffered cohesive failure

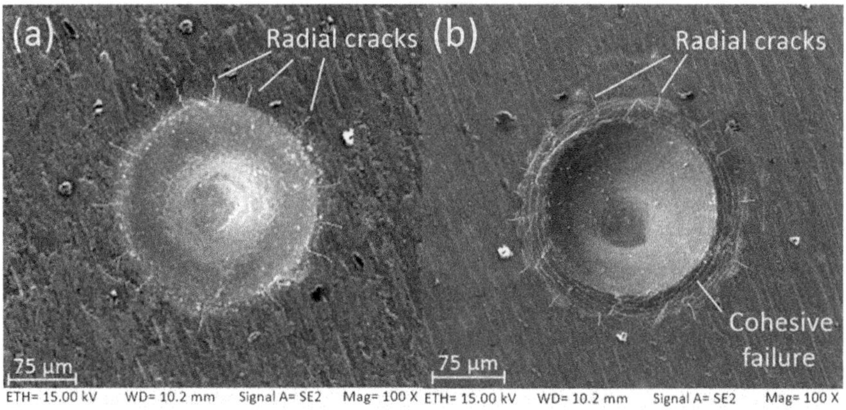

FIGURE 10.5 Rockwell-C indentation impression on (a) AlCrSiN and (b) AlCrN thin film.

around the impression edge along with radial cracks which can be categorized into HF3 or HF4 quality rankings. Interestingly, both the films did not showcase any formation of black ellipse at the impression edge (or away from it) indicating no peeling of the film material. Also, coating-substrate adhesion till HF4 quality ranking is acceptable [40]. Vopát et al. [40] deposited AlCrSiN (2 at% Si) using PVD technique on cemented carbide and observed better adhesion (HF2) with films deposited with −100 V bias voltage compared to −80 V and −120 V. This could be attributed to better cubic CrN morphology with amorphous Si_xN_y phases and absence of AlN phases inducing better H/E and H^3/E^2 factors.

10.4 CONCLUSIONS

AlCrN and AlCrSiN films were synthesized on chrome-nitrided (CrN interlayer) medical-grade 316 LVM stainless steel using cathodic arc deposition technique. The topography, morphology, tribo-behavior (lubricated), and biological response were comprehensively studied. The following conclusions can be drawn from the work:

a) Introduction of Si in AlCrN resulted in slight widening of the diffraction lines in the XRD analysis, indicating a decrease in crystallite size and lattice parameter. No peaks associated with fcc/hcp/wurtzite AlN were recorded, suggesting a more stabilized fcc structural makeup of the film which is also corroborated with Raman bands of the films.
b) Si-incorporated AlCrN film had responded with improved hardness and elastic modulus propounding to higher H/E and H^3/E^2 factors attributed to the formation of amorphous SixNy phases inducing grain refinement causing hindrance in the grain boundary. Also, AlCrSiN film had recorded better scratch hardness due to smaller width of the nanoscratch wear track owing to improved resistance to deformation than AlCrN film. Further, both the films have showcased appreciable adhesion accessed using Rockwell-C

indentation test, wherein AlCrSiN film has suffered smaller radial cracks, and AlCrN film has witnessed cohesive failure along with radial cracks.

ACKNOWLEDGMENT

The authors would like to thank Ingenious Tools and Design Solution Pvt. Ltd. Ghaziabad for providing assistance with thin film deposition and Central Research Facility, Indian Institute of Technology Kharagpur, Centre for Nanoscience and Nanotechnology, Sathyabama Institute of Science and Technology, Chennai, for conducting the experiments pertaining to the characterization of the films.

REFERENCES

1. Mehr, Ali Kosari, Mohammad Reza Zamani Meymian, and Abbas Kosari Mehr. "Nanoindentation and nanoscratch studies of submicron nanostructured Ti/TiCrN bilayer films deposited by RF-DC co-sputtering method." *Ceramics International* 44, no. 17 (2018): 21825–21834.
2. Falsafein, Maedeh, Fakhreddin Ashrafizadeh, and Alireza Kheirandish. "Influence of thickness on adhesion of nanostructured multilayer CrN/CrAlN coatings to stainless steel substrate." *Surfaces and Interfaces* 13 (2018): 178–185.
3. Li, Hang, Zhaoli Liu, Jianliang Li, Jiewen Huang, Jian Kong, Qiujie Wu, and Dangsheng Xiong. "Effects of Hf addition on the structure, mechanical and tribological properties of CrN film." *Surface and Coatings Technology* 397 (2020): 126067.
4. Huang, Kai, Xueqian Cao, Linggang Kong, Zhibin Lu, Qi Ding, and Haitao Hu. "Effect of Ag content on friction and wear properties of Ag and V co-doped CrN coatings at 25–700 C." *Ceramics International* 47, no. 24 (2021): 35021–35028.
5. Filho, Luimar, Susann Schmidt, Klaus Leifer, Håkan Engqvist, Hans Högberg, and Cecilia Persson. "Towards functional silicon nitride coatings for joint replacements." *Coatings* 9, no. 2 (2019): 73.
6. Onoprienko, A. A., V. I. Ivashchenko, P. L. Scrynskyy, A. K. Sinelnichenko, A. A. Kozak, A. M. Kovalchenko, and E. I. Olifan. "Influence of Nitrogen on the Microstructure, Hardness, and Tribological Properties of Cr–Ni–B–C–N Films Deposited by DC Magnetron Sputtering." *Journal of Superhard Materials* 42 (2020): 68–77.
7. Du, Jian W., Li Chen, Jie Chen, and Chun Hu. "Influence of oxygen addition on the structure, mechanical and thermal properties of CrN coating." *Surface and Coatings Technology* 411 (2021): 126992.
8. Xu, Xing, Jianfang Sun, Zhibiao Xu, Zhujun Li, and Fenghua Su. "Microstructure, electrochemical and tribocorrosion behaviors of CrCN nanocomposite coating with various carbon content." *Surface and Coatings Technology* 411 (2021): 126997.
9. Chen, Yu, Zhijian Zhang, Tiechui Yuan, Fangsheng Mei, Xiaoliang Lin, Jiangxiong Gao, Weifei Chen, and Yunbao Xu. "The synergy of V and Si on the microstructure, tribological and oxidation properties of AlCrN based coatings." *Surface and Coatings Technology* 412 (2021): 127082.
10. Kumar, Sunil, Saikat Ranjan Maity, and Lokeswar Patnaik. "Friction and tribological behavior of bare nitrided, TiAlN and AlCrN coated MDC-K hot work tool steel." *Ceramics International* 46, no. 11 (2020): 17280–17294.
11. Reiter, A. E., V. H. Derflinger, B. Hanselmann, T. Bachmann, and B. Sartory. "Investigation of the properties of Al1– xCrxN coatings prepared by cathodic arc evaporation." *Surface and Coatings Technology* 200, no. 7 (2005): 2114–2122.

12. Warcholinski, B., A. Gilewicz, P. Myslinski, E. Dobruchowska, D. Murzynski, P. Kochmanski, Krzysztof Rokosz, and Steinar Raaen. "Effect of nitrogen pressure and substrate bias voltage on the properties of Al–Cr–B–N coatings deposited using cathodic arc evaporation." *Tribology International* 154 (2021): 106744.
13. Sabitzer, C., J. Paulitsch, S. Kolozsvári, R. Rachbauer, and P. H. Mayrhofer. "Influence of bias potential and layer arrangement on structure and mechanical properties of arc evaporated Al–Cr–N coatings." *Vacuum* 106 (2014): 49–52.
14. Fuentes, G. G., R. Rodriguez, J. C. Avelar-Batista, J. Housden, F. Montalá, L. J. Carreras, Ana Belén Cristóbal, J. J. Damborenea, and T. J. Tate. "Recent advances in the chromium nitride PVD process for forming and machining surface protection." *Journal of Materials Processing Technology* 167, no. 2–3 (2005): 415–421.
15. Jäger, Nikolaus, Michael Meindlhumer, S. Spor, Hynek Hruby, J. Julin, Andreas Stark, F. Nahif, Julius Keckes, Christian Mitterer, and Rostislav Daniel. "Microstructural evolution and thermal stability of AlCr (Si) N hard coatings revealed by in-situ high-temperature high-energy grazing incidence transmission X-ray diffraction." *Acta Materialia* 186 (2020): 545–554.
16. Sribalaji, M., OS Asiq Rahman, Tapas Laha, and Anup Kumar Keshri. "Nanoindentation and nanoscratch behavior of electroless deposited nickel-phosphorous coating." *Materials Chemistry and Physics* 177 (2016): 220–228.
17. Lee, Jyh-Wei, and Yue-Chyuan Chang. "A study on the microstructures and mechanical properties of pulsed DC reactive magnetron sputtered Cr–Si–N nanocomposite coatings." *Surface and Coatings Technology* 202, no. 4–7 (2007): 831–836.
18. Endrino, J. L., S. Palacín, M. H. Aguirre, A. Gutiérrez, and F. Schäfers. "Determination of the local environment of silicon and the microstructure of quaternary CrAl (Si) N films." *Acta Materialia* 55, no. 6 (2007): 2129–2135.
19. He, Linqing, Li Chen, and Yuxiang Xu. "Interfacial structure, mechanical properties and thermal stability of CrAlSiN/CrAlN multilayer coatings." *Materials Characterization* 125 (2017): 1–6.
20. Wang, Yu Xi, Sam Zhang, Jyh-Wei Lee, Wen Siang Lew, Deen Sun, and Bo Li. "Toward hard yet tough CrAlSiN coatings via compositional grading." *Surface and Coatings Technology* 231 (2013): 346–352.
21. Chang, Chun-Chi, Hsien-Wei Chen, Jyh-Wei Lee, and Jenq-Gong Duh. "Influence of Si contents on tribological characteristics of CrAlSiN nanocomposite coatings." *Thin Solid Films* 584 (2015): 46–51.
22. Liu, Yanmei, Tie-Gang Wang, Wei Lin, Qiang Zhu, Bing Yan, and Xiang Hou. "Microstructure and properties of the AlCrSi (O) N tool coatings by arc ion plating." *Coatings* 10, no. 9 (2020): 841.
23. Warcholinski, Bogdan, Adam Gilewicz, Piotr Myslinski, Ewa Dobruchowska, Dawid Murzynski, and Tatyana A. Kuznetsova. "Effect of silicon concentration on the properties of Al-Cr-Si-N coatings deposited using cathodic arc evaporation." *Materials* 13, no. 21 (2020): 4717.
24. Chen, Yun, Hao Du, Ming Chen, Jun Yang, Ji Xiong, and Haibo Zhao. "Structure and wear behavior of AlCrSiN-based coatings." *Applied surface science* 370 (2016): 176–183.
25. Tritremmel, Christian, Rostislav Daniel, M. Lechthaler, P. Polcik, and Christian Mitterer. "Influence of Al and Si content on structure and mechanical properties of arc evaporated Al–Cr–Si–N thin films." *Thin Solid Films* 534 (2013): 403–409.
26. Soldán, J., Jörg Neidhardt, B. Sartory, R. Kaindl, R. Čerstvý, Paul Heinz Mayrhofer, R. Tessadri, P. Polcik, M. Lechthaler, and Christian Mitterer. "Structure–property relations of arc-evaporated Al–Cr–Si–N coatings." *Surface and Coatings Technology* 202, no. 15 (2008): 3555–3562.

27. Kaindl, R., Robert Franz, J. Soldan, A. Reiter, Peter Polcik, Christian Mitterer, Bernhard Sartory, Richard Tessadri, and M. O'Sullivan. "Structural investigations of aluminum-chromium-nitride hard coatings by Raman micro-spectroscopy." *Thin Solid Films* 515, no. 4 (2006): 2197–2202.
28. Vepřek, Stan. "The search for novel, superhard materials." *Journal of Vacuum Science & Technology A: Vacuum, Surfaces, and Films* 17, no. 5 (1999): 2401–2420.
29. Patnaik, Lokeswar, Saikat Ranjan Maity, and Sunil Kumar. "Evaluation of crack resistance and adhesive energy of AlCrN and Ag doped a-C films deposited on chrome nitrided 316 LVM stainless steel." *Advances in Materials and Processing Technologies* 8, no. sup3 (2022): 1048–1069.
30. Zhao, Biaochun, Xiaoxiao Zhao, Liangliang Lin, and Lingli Zou. "Effect of bias voltage on mechanical properties, milling performance and thermal crack propagation of cathodic arc ion-plated TiAlN coatings." *Thin Solid Films* 708 (2020): 138116.
31. Park, In-Wook, Dong Shik Kang, John J. Moore, Sik Chol Kwon, Jong Joo Rha, and Kwang Ho Kim. "Microstructures, mechanical properties, and tribological behaviors of Cr–Al–N, Cr–Si–N, and Cr–Al–Si–N coatings by a hybrid coating system." *Surface and Coatings Technology* 201, no. 9–11 (2007): 5223–5227.
32. Hirai, Makoto, Yasuhiro Ueno, Tsuneo Suzuki, Weihua Jiang, Constantin Grigoriu, and Kiyoshi Yatsui. "Characteristics of (Cr1-x, Alx) N films prepared by pulsed laser deposition." *Japanese Journal of Applied Physics* 40, no. 2S (2001): 1056.
33. Kawate, Masahiro, Ayako Kimura, and Tetsuya Suzuki. "Microhardness and lattice parameter of Cr 1− x Al x N films." *Journal of Vacuum Science & Technology A: Vacuum, Surfaces, and Films* 20, no. 2 (2002): 569–571.
34. Schiøtz, Jakob, Francesco D. Di Tolla, and Karsten W. Jacobsen. "Softening of nanocrystalline metals at very small grain sizes." *Nature* 391, no. 6667 (1998): 561–563.
35. Ding, Xing-zhao, X. T. Zeng, and Y. C. Liu. "Structure and properties of CrAlSiN Nanocomposite coatings deposited by lateral rotating cathod arc." *Thin Solid Films* 519, no. 6 (2011): 1894–1900.
36. Chen, Hsien-Wei, Yu-Chen Chan, Jyh-Wei Lee, and Jenq-Gong Duh. "Oxidation behavior of Si-doped nanocomposite CrAlSiN coatings." *Surface and Coatings Technology* 205, no. 5 (2010): 1189–1194.
37. Dey, Dipanjan, Kalinga Simant Bal, Anitesh Kumar Singh, and Asimava Roy Choudhury. "Hardness and wear behaviour of multiple component coating on Ti-6Al-4V substrate by laser application." *Optik* 202 (2020): 163555.
38. Chen, Wanglin, Daoda Zhang, Dengcan Yao, Shihong Zhang, and Weiwei Wu. "Investigations on microstructure and mechanical properties of containing-Si coatings." *Surface Engineering* 33, no. 7 (2017): 536–541.
39. Patnaik, Lokeswar, Saikat Ranjan Maity, and Sunil Kumar. "Mechanical and tribological assessment of composite AlCrN or a-C: Ag-based thin films for implant application." *Ceramics International* 47, no. 5 (2021): 6736–6752.
40. Vopát, Tomáš, Martin Sahul, Marián Haršáni, Ondřej Vortel, and Tomáš Zlámal. "The tool life and coating-substrate adhesion of AlCrSiN-coated carbide cutting tools prepared by LARC with respect to the edge preparation and surface finishing." *Micromachines* 11, no. 2 (2020): 166.

11 Anti-wear and Bacterial Adhesion Properties of Si-Doped AlCrN Thin Films for Biomedical Application

Lokeswar Patnaik and Magdalena Łępicka

11.1 INTRODUCTION

AlCrN-based coatings deposited by physical vapor deposition (PVD) exhibit high hardness (27–32 GPa), excellent thermal stability up to 800–900°C, and favorable tribological performance, which have underpinned their extensive application in industrial mold and tool protection [1]. Nevertheless, dry, high-speed machining continues to pose substantial challenges to the durability of AlCrN coatings. Researchers discovered that adding Si to AlCrN-fabricated AlCrSiN coatings made of the amorphous Si_3N_4 phase is a useful way to enhance microstructure and properties, such as improved tribology behaviour, longer service life, higher hardness (35–50 GPa), and higher thermal stability (1000–1100°C) [1]. AlCrSiN coatings with a high Si concentration (5–10 wt%) showed reduced adhesion strength and increased brittleness, which were caused by the intrinsic flaws in nitride coatings [2]. Due to significant variations in the chemical compositions, mechanical characteristics, and coefficient of thermal expansion between the coatings and substrate, they are especially easily peeled off under coupled mechanical and thermal loading [3]. Strategies such as gradient architectures [4], multilayer designs, multicomponent compositions, and grain-refinement techniques have been demonstrated to alleviate these mismatches and enhance the adhesion and toughness of AlCrSiN coatings.

Soldán et al. [5] found this phenomenon more pronounced with AlCrSiN films having Si at% ≥ 2 due to deterioration in crystallinity. Bogdan et al. [6] found that AlCrSiN films containing up to 2 at% Si exhibit superior mechanical properties compared to AlCrN. However, further Si enrichment increases the crystallite size while reducing the lattice parameter, a combination that ultimately undermines the coating's mechanical performance. AlCrN coatings have been investigated for biomedical applications [7], and silicon-based modifications have likewise been extensively explored in this field [8]. Many authors have demonstrated the dry sliding behavior of AlCrN [7, 9] and AlCrSiN [10] films; however, evaluation of these films

DOI: 10.1201/9781003750659-11

in lubricated sliding [11] and studies on antibacterial adhesion are seldom performed. In the present work, AlCrN and Si-incorporated AlCrN that is, AlCrSiN were deposited on chrome-nitrided (CrN) medical grade 316 LVM stainless steel using cathodic arc deposition technique. The tribological behavior of these coatings was evaluated under lubrication in simulated body fluid, using a Ti6Al4V counterface to replicate a metal-on-metal acetabular articulation. Previous studies have confirmed the suitability of titanium and its alloys for acetabular cup applications in hip implants [12, 13]. Consequently, as a compliment to these analyses, post lubricated tribotest, the worn surface of these films was again subjected to Raman studies, nanoindentation, and nanoscratch tests to understand the effect of lubricated sliding on their mechanical properties and structural integrity. Lastly, the bacterial growth (*Escherichia coli* and *Bacillus subtilis*) on as-deposited (as control) and worn coated surfaces was studied to evaluate the bacterial adhesion on these films.

11.2 MATERIALS AND METHOD

Deposition was carried out on medical-grade 316 LVM stainless steel discs (ϕ 30 mm × 5 mm) via cathodic arc evaporation (Ingenious Tools and Design Solution, India). Prior to coating, substrates were sequentially ground (abrasive papers, 200–1000 grit), polished with diamond paste, ultrasonically cleaned in alkaline solution, rinsed with deionized water, and vacuum-dried. The polished 316 LVM exhibited a mean surface roughness of 0.60 ± 0.11 μm, hardness of 3.2 ± 0.8 GPa, and elastic modulus of 198 ± 18 GPa. A CrN adhesion interlayer (~1 μm) was first deposited from a 99.99%-pure Cr cathode: the chamber was evacuated to 4×10^{-3} Pa, heated to 380 °C, backfilled with Ar (30 sccm) to 2×10^{-2} Pa, then with N_2 (800 sccm) at 2 N pressure; the substrate bias was –80 V and arc current 90 A for 30 minutes at 4.1×10^{-2} Pa. After deposition, the CrN-nitrided substrates were lightly polished and returned to the chamber for top-coat growth. Using $Al_{70}Cr_{30}$ and $Al_{68}Cr_{30}Si_2$ cathodes, AlCrN and AlCrSiN films (~1.5 μm) were deposited at 400 °C, 1×10^{-3} Pa, with an Ar-etch (15 min at –200 V), then a 100 A arc current and –100 V substrate bias under 4 Pa N_2 (monitored by capacitive gauges) for 45 minutes. Coating morphology and composition were examined by SEM (Zeiss Supra 55 VP) coupled with energy-dispersive spectroscopy. The simulated tribotests in lubricating condition were performed on a pin-on-disc tribometer (TRB^3, Anton Paar, India) against Ti6Al4V alloy counterbody (ϕ8 mm × 30 mm). The relevance of Ti6Al4V for acetabular cup material is described elsewhere [13]. The sample preparation was done following ASTM G99 standard. The lubricated tribotests were performed at 5 N load with 0.1 m/s sliding velocity for 20,000 cycles at 1 Hz frequency generating a Hertzian contact pressure of 1 GPa. The ambience of the tribotest chamber was kept at 37±1 °C maintaining a relative humidity of 50%. The friction data was recorded using the ratio of average normal load force and tangential force by virtue of strain gauges after filtering the data (disturbance or noises) using low-pass filter of 250 Hz. The parameters mentioned for the simulated tribotest pertain to wear analysis of hip implant materials using tribotest [14–16]. The phase and polymorphic analysis of the coatings was conducted using Raman spectroscopy integrated with He-Ne laser (λ= 632.8 nm) focused with a numerical aperture of 0.5 (HR800, Horiba Jobin Vyon, Japan).

The governing equations for obtaining the particulars of the nanoindentation and nanoscratch test following Oliver and Pharr method are categorically presented in a study by Sribalaji et al. [17]. The particulars of the ion concentration (in g/1,000 ml) in the simulated body fluid are NaCl (8.035), $NaHCO_3$ (0.355), KCl (0.225), $K_2HPO_4.3H_2O$ (0.231), $MgCl_2.H_2O$ (0.311), 1 M HCl (40 ml), $CaCl_2$ (0.292), Na_2SO_4 (0.072), $((CH_2OH)3CNH_2$ (6.118), and HCl (appropriate pH adjusted to 7.4) [18]. The specific wear rate (K in mm^3/Nm) of the coated substrate and pin was calculated following Archard's equation by measuring the wear volume loss (mm^3) upon load applied (N) and sliding distance (m) [19]. The mass loss in the materials was measured using electronic weighing machine (P-7 B, 220 G, Mxrady lab, India) having an accuracy of 10^{-3} g. The 2D roughness measurement was conducted on the worn-out coated surfaces using profilometer (Dektak XT, Bruker, India). The microhardness of the worn-out surfaces was obtained using Vickers microhardness tester (HM 200 System D, Mitutoyo, India) with a load of 200 g keeping a dwell time of 5 s. Nanoindentation and nanoscratch test were further conducted to evaluate the effect of lubricating simulated tribotest on the mechanical properties of the worn coated surface. The worn surface of the coatings was observed under SEM to understand the underlying wear mechanism. What is more, to understand the structural change (if any) in the coatings subjected to lubricated tribotest, Raman spectroscopy was performed. The antibacterial assessment of the coatings was done against gram-negative (*Escherichia coli*) and gram-positive (*Bacillus subtilis*) bacteria based on JIS Z 2801:2010 standard [20]. The strains (single pure colony) were cultured in LB medium (Luria Bertani) incubated for 18 h at 37 °C shaking at 150 rpm (NXT-127CI, Next Era Technologies, India). The bacterial cells were harvested by means of centrifugation at 5,000 rpm for 15 minutes and subsequently washed with a buffer solution (sterile phosphate). The previously sterilized uncoated and coated substrates were shaken at 200 rpm at 37 °C with 1 mL of harvested bacterial suspension in a conical flask, and the number of surviving bacteria was counted after 12 h and 24 h via surface spread plate method. Subsequently, 1 mL of the sample (separate for uncoated substate, AlCrN and AlCrSiN) was taken and subjected to serial dilution using saline solution. Finally, 1 mL of the diluent (of each sample) was spread on an LB agar plate and incubated for 24 h at 37 °C. Three replicates of each sample were tested for repeatability. The inhibition efficiency (%) was calculated using the average colony forming unit (CFUs) of the uncoated substrate and coated substrate (S) following Equation 1, [21] where *No* and *N* are the average CFUs on as deposited and worn coated surfaces respectively.

$$\text{Inhibition efficiency rates}(\%) = \frac{N_o - N}{N_o} \times 100 \tag{1}$$

11.3 RESULTS AND DISCUSSION

11.3.1 Wear Behavior

For AlCrN, the friction trace initially stabilized but then displayed periodic fluctuations, reflecting successive cycles of transfer-layer formation and breakdown. In

TABLE 11.1
Wear test results and follow-up analysis

Tests	Particulars	Thin Films: AlCrN	Thin Films: AlCrSiN	p-Value	Tukey's Grouping
Wear test	Friction coefficient	0.38±0.11	0.31±0.09	>0.05	S
	Specific wear rate of the substrate (mm^3/N.m)	(9.87±2.3) ×10^{-8}	(2.54±1.8)×10^{-8}	<0.05	D
	Specific wear rate of the counterbody (mm^3/N.m)	(10.1±1.9)×10^{-6}	(7.9±1.6)×10^{-7}	<0.05	D
Worn surface analysis	Width of the wear track on the coated surface (in mm)	2.4±0.5	1.8±0.4	>0.05	S
	Worn coated surface roughness (in µm)	4.9±1.3	3.3±1.2	>0.05	S
	Microhardness of the worn coated surface (in Hv)	2846±411	4089±325	<0.05	D

D: Different Tukey's group indicating statistical difference
S: Same Tukey's group indicating no statistical difference

contrast, AlCrSiN exhibited a run-in period up to approximately 17,575 cycles, after which the coefficient of friction remained constant for the remainder of the wear test. The results of the lubricated wear test and follow-up analysis of the worn surfaces are given in Table 11.1. Nonetheless, AlCrSiN exhibited a lower mean friction coefficient than AlCrN. Silicon incorporation promotes the formation of protective tribofilms during sliding – a behavior similarly documented by Lee and Elizer in Fe-Ni-Si alloys [22]. The wear depth profile of the wear scar on AlCrSiN (821 nm) was relatively lower than AlCrN (1,500 nm) owing to adhesion-induced higher friction coefficient propounding to higher specific wear rate in the latter. In fact, both the films were not completely delaminated during the lubricated sliding. CrN as an interlayer acted as a barrier for the Ti6Al4V counterbody to completely penetrate through the films. The wear scar profile of AlCrN (see Figure 11.1(b)) demonstrated plastic deformation at the edge of the wear track along with material buildup at places in the wear crater, indicating plastic deformation-induced material transfer and adhesion. On the other hand, the wear scar profile of AlCrSiN showed furrows at the bottom of the wear crater owing to high brittleness (high H/E).

Further, the specific wear rate of Ti6Al4V counterbody registered with AlCrN was one order higher than AlCrSiN, indicating severe counterbody material loss during sliding against AlCrN. The SEM micrograph of the worn Ti6Al4V counterbody slid against AlCrN and AlCrSiN coated surface along with EDS spectrum is presented in Figure 11.1(c-c′) and (d-d′) respectively. Also, it is evident that material transfer from Ti6Al4V counterbody was higher against AlCrN compared to AlCrSiN owing to larger diameter of the worn counterface when slid against the former. The adhesive patches on the counterbody (see Figure 11.1(c) are suggestive of plastic deformation during sliding of Ti6Al4V-AlCrN tribopair. One of the inherent problems of

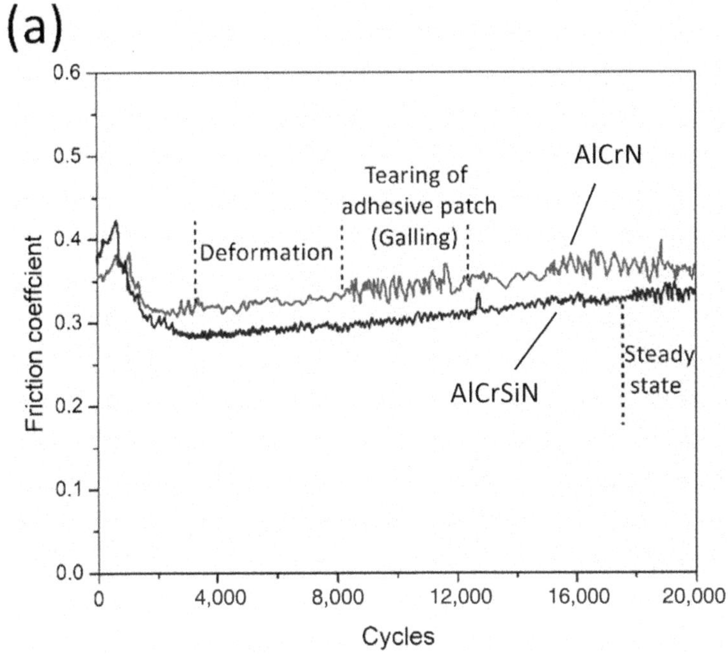

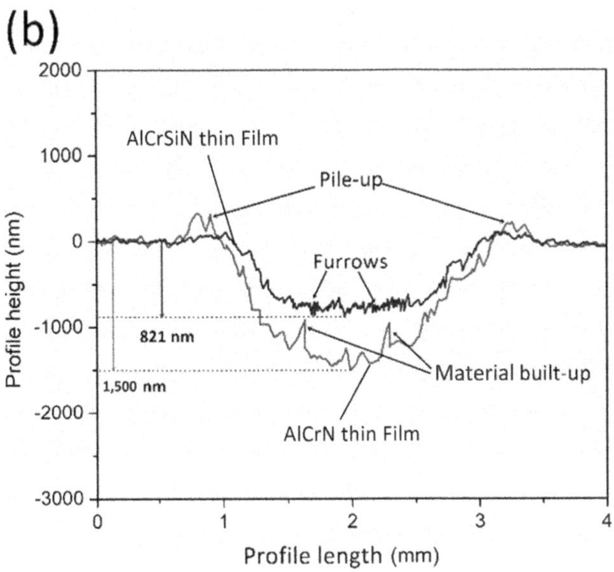

FIGURE 11.1 (a) Friction dynamics recorded during wear test (b) wear scar profile on the coated surfaces; SEM micrograph and corresponding point EDS spectrum of Ti6Al4V worn pin slid against (c-c′) AlCrN and (d-d′) AlCrSiN.

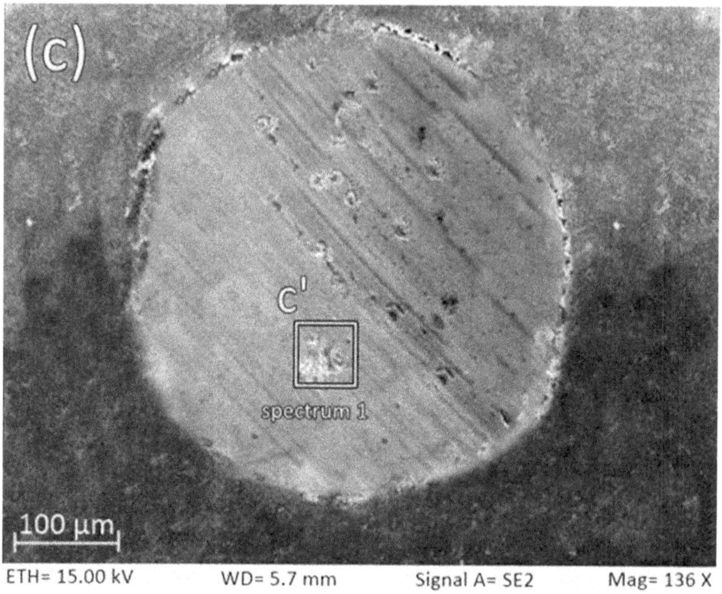

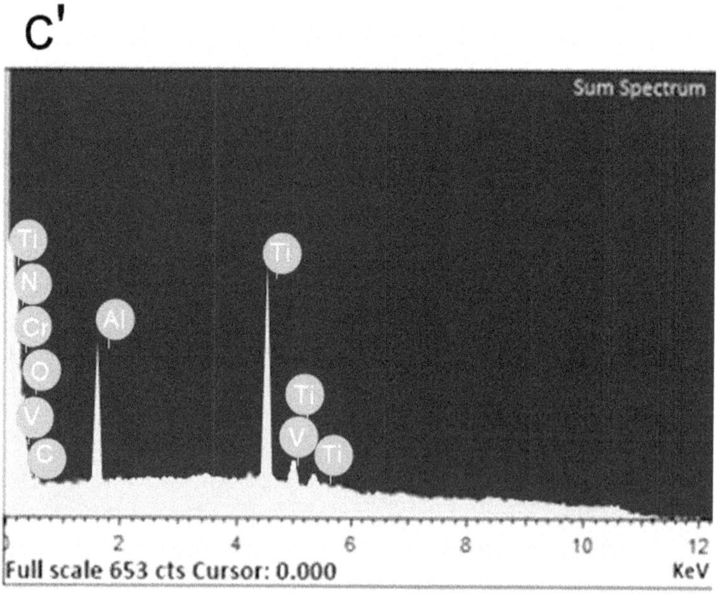

FIGURE 11.1 (Continued)

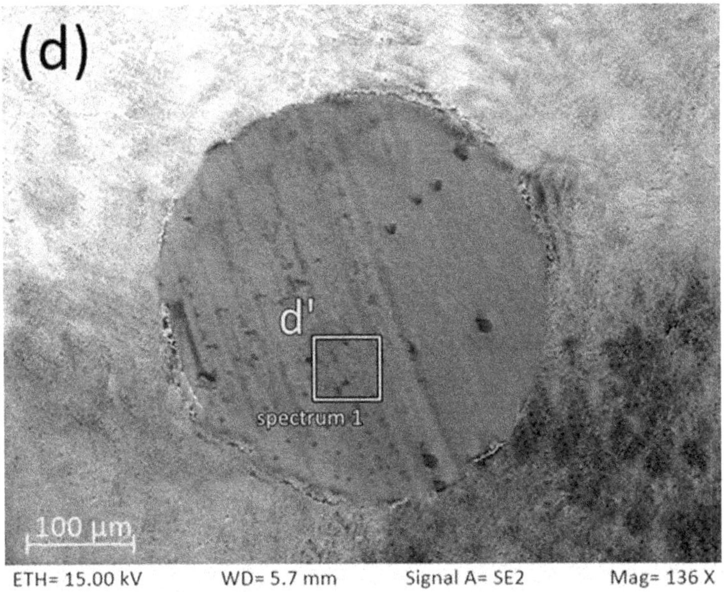

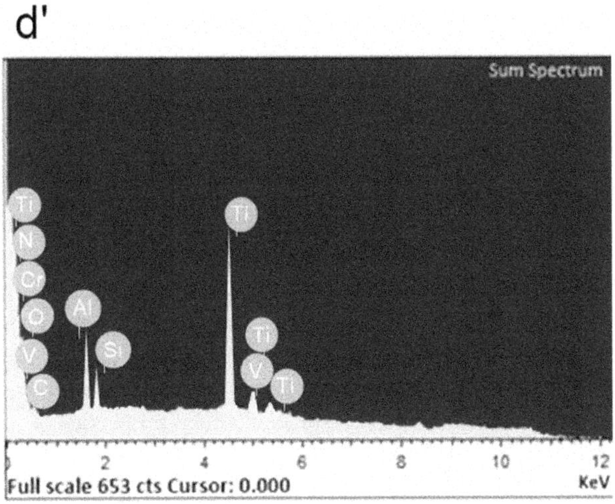

FIGURE 11.1 (Continued)

Ti6Al4V is low shear strength owing to its HCP structure, resulting in severe scuffing which led to its degrading and sticking as an adhesive layer on the film.

The SEM micrographs of the worn coated surfaces and their corresponding point EDS data are presented in Figure 11.2. The AlCrN worn coated surface (see Figure 11.2(a) and (b)) demonstrated the formation of visible adhesive layer pointing toward galling and cyclic upsetting leading to adhesive smearing during the simulated tribotest. There was evidence of plowing on the adhesive tribolayer as well. The phenomenon related to material transfer, plastic deformation and adhesion in AlCrN could be attributed to the chemical affinity of Al and Ti toward each other as reported in literature [11, 23]. Previously, it was established that AlCrN (38.1 at%) has higher Al content than AlCrSiN (33.4 at%) attributed to Si introduction depleting the Al sites. Therefore, the tribochemical reaction against Ti-alloy could have been higher against AlCrN due to high Al content present in it. Meanwhile, AlCrSiN worn surface was clearly pointing toward fatigue-induced cracking followed by micro-abrasion. From the corresponding point EDS spectrum of the worn surfaces of AlCrN and AlCrSiN shown in Figure 11.2(b′) and (d′) respectively, it is quite clear that strong intensity peaks of the Ti6Al4V counterbody material in the EDS spectrum of worn AlCrN indicate severe material transfer. Nonetheless, it is important to note that the EDS spectrum showed no evidence of substrate elements such as Fe, Ni, Mo among others suggesting the underneath substrate was not exposed during lubricated sliding. Lee and Elizer [22] had demonstrated segregation of silicon and suggested that it has the tendency to diffuse to the counterface attributed to temperature generation at the sliding interface. Also, with the advent of surface deformation-induced lattice defect, the ability of silicon diffusion increases. The evidence of silicon on Ti6Al4V counterface is evident from the point EDS spectrum shown in Figure 11.1(d). Further, the movement of silicon toward the sliding interface results in oxidation under localized temperature and pressure during sliding. In another work, Chen et al. [24] had reported higher oxidizing capacity of Si_xN_y over c-AlCrN phase. They had also suggested that Ti and Al tribo-chemical reaction leading to the generation of thick and loose adhesive film which is evident from the adhesive layer formed over the worn AlCrN surface shown in Figure 11.2(a-a′). Over the period of sliding, the adhered patch gets deformed and scrapped away resulting in adhesive wear or galling (see Figure 11.2(a-b)) evident from the high friction dynamics of AlCrN compared to AlCrSiN (see Figure 11.1(a)). On the other hand, in case of AlCrSiN, the oxidized Si_xN_y acts as a protective hard interface leading to relatively lower friction coefficient. In fact, a higher H^3/E^2 factor (0.40±0.11) in AlCrSiN restricts deformation leading to the film witnessing fatigue-induced mild abrasion leading to the formation of furrows at the bottom of the wear scar (see Figure 11.1(b)). Nonetheless, it would be interesting to understand the change in the structural makeup and deterioration of the mechanical properties of these films post lubricated sliding.

11.3.2 Structural Makeup, Mechanical Properties, and Antibacterial Activity of the Worn Coated Surfaces

Raman pattern of the worn AlCrN and AlCrSiN surfaces presented in Figure 11.3 demonstrates the change in structural integrity of the films after lubricated sliding. It can be seen from the wear scar profile (see Figure 11.1(b)) that the depth of the wear scar in AlCrN was 1,500 nm (or 1.5 μm) indicating severe deterioration of the AlCrN topcoat exposing the CrN interlayer. The Raman spectrum of the worn AlCrN

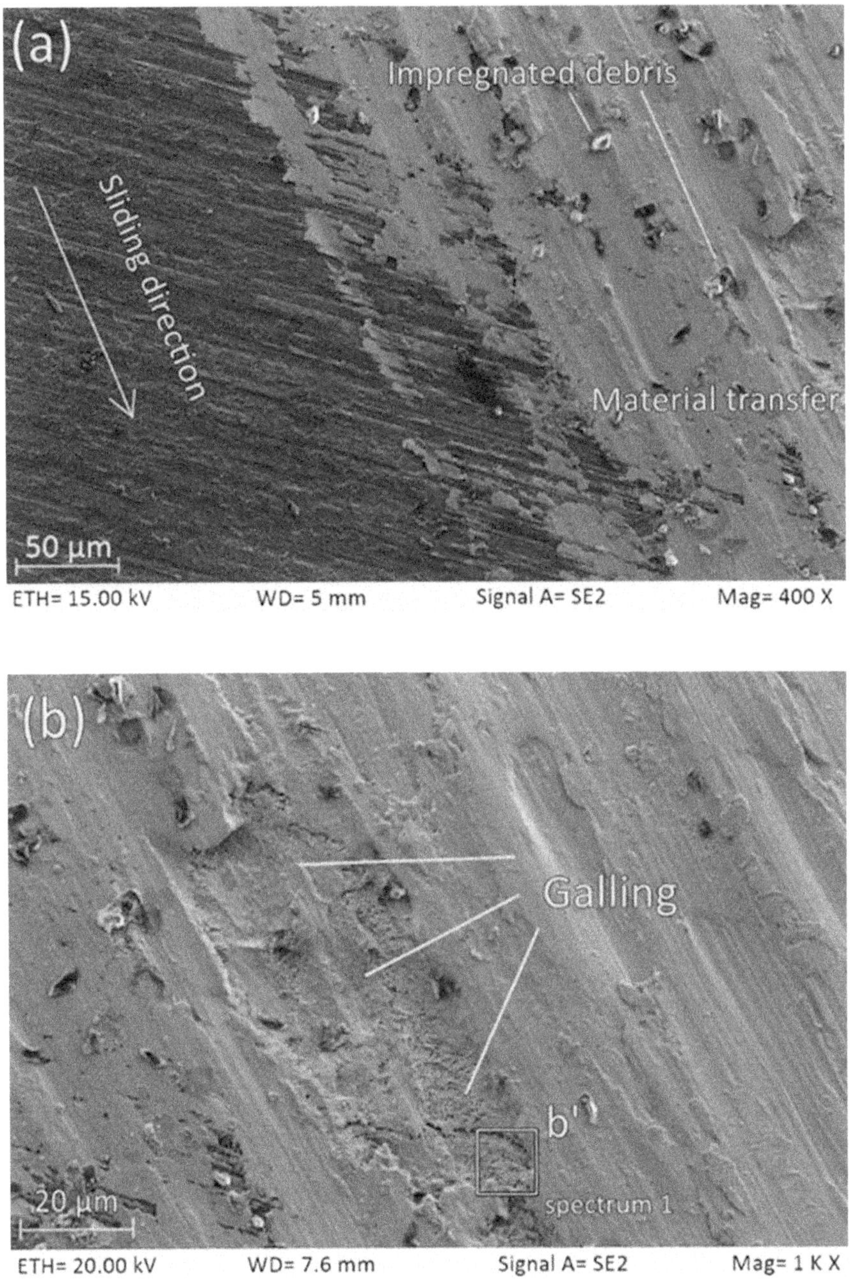

FIGURE 11.2 Worn surface SEM micrographs and corresponding EDS spectrum of (a, b-b′) AlCrN and (c, d-d′) AlCrSiN.

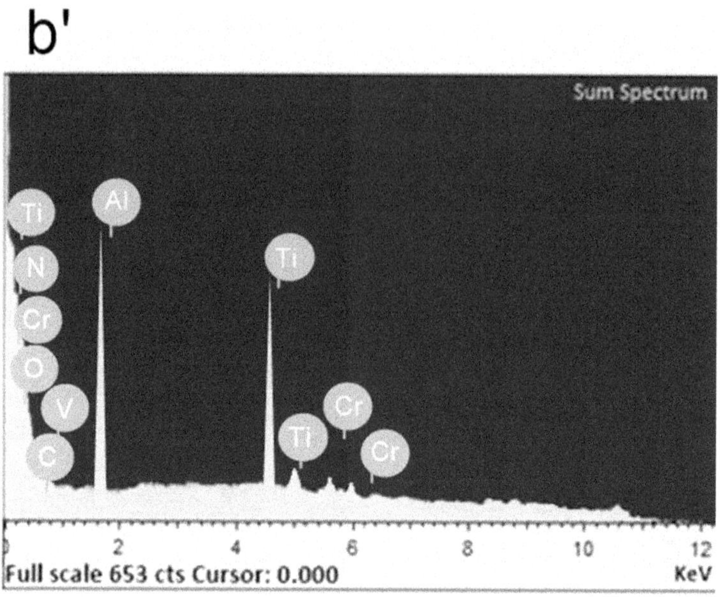

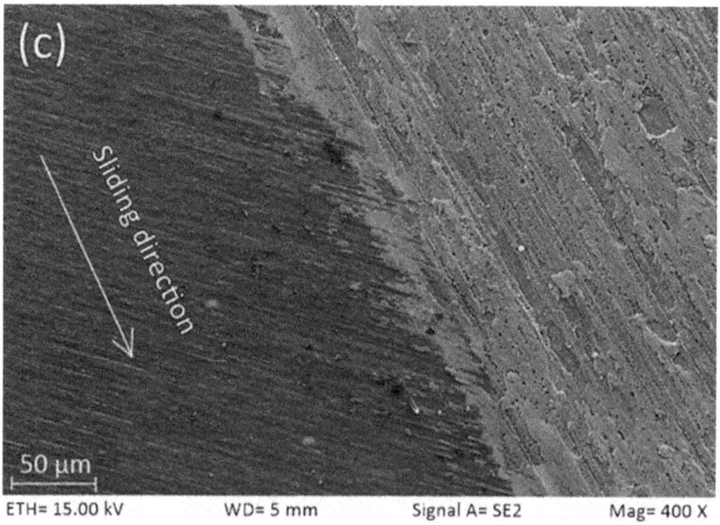

FIGURE 11.2 (Continued)

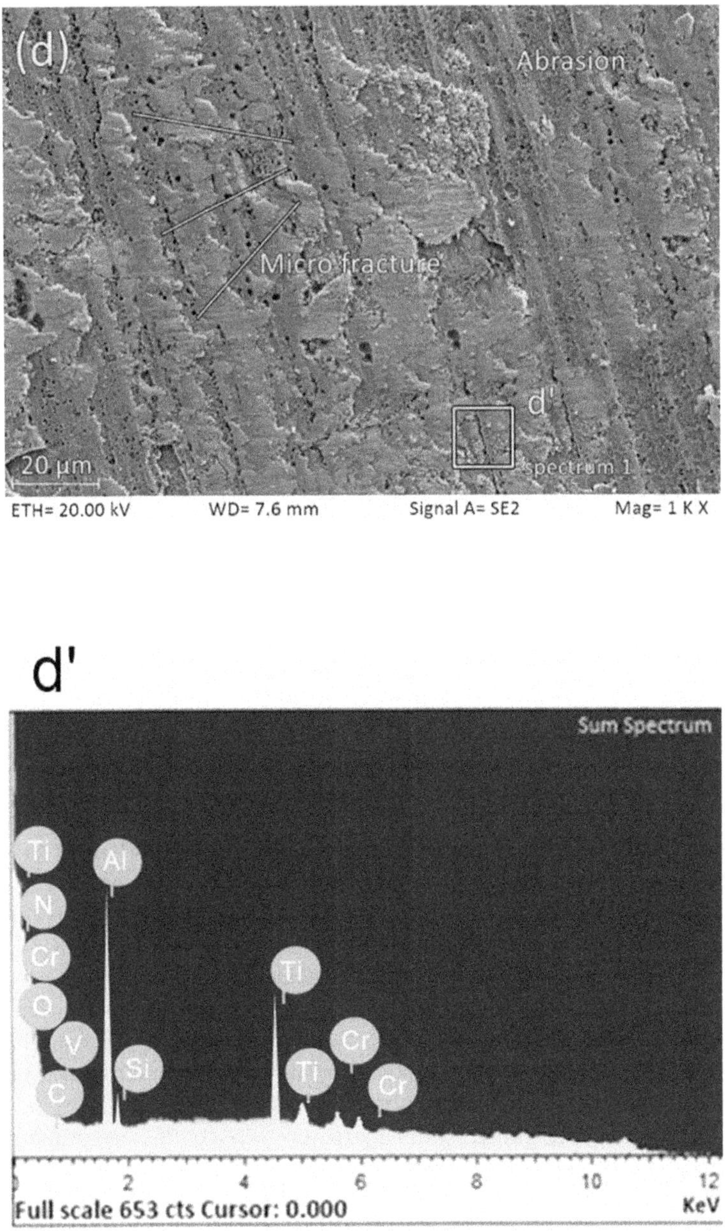

FIGURE 11.2 (Continued)

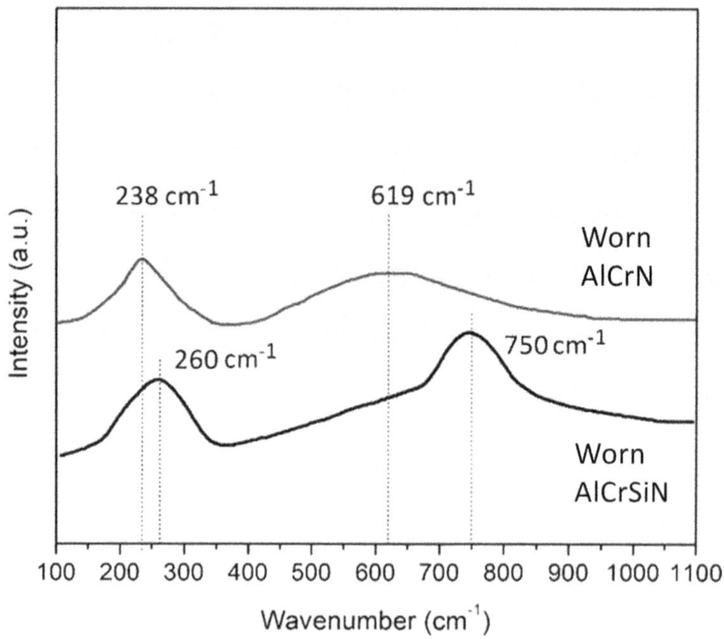

FIGURE 11.3 Raman spectra of the worn coated surfaces.

coating had shifted to lower and weaker bands at 238 cm^{-1} and 619 cm^{-1}, indicating longitudinal and transverse acoustic mode of c-CrN phase pertaining to the CrN interlayer. Interestingly, no peaks ascribed to anatase or rutile phase of TiO_2 was detected between 200 cm^{-1} and 600 cm^{-1} spectrum, suggesting loose and amorphous nature of the adhesive layer (from Ti6Al4V counterbody) on the worn AlCrN surface. Further, the spectrum did not demonstrate any peaks corresponding to the crystalline phase of Al_2O_3 [25], Cr_2O_3 [26], and V_2O_5 [27], suggestive of amorphous phase of the adhesive patch on the AlCrN worn surface. On the other hand, the structural integrity of the worn AlCrSiN film was intact post tribotest owing to lower wear. The typical temperature requirement for the formation of potential oxides of Ti [28], Al [29], Cr [26], V [30], and Si [31] with crystalline structure is > 400 °C, one of the plausible reasons for the inhibition of crystalline oxide formation could be the continuous flow of the lubricant taking away the heat generated during sliding.

In summary, the worn surface of AlCrN showed the formation of an adhesive layer, whereas AlCrSiN witnessed fatigue-induced micro-abrasion. Since both the films had demonstrated significantly different nature of wear mechanism, it is important to understand the response of the lubricated wear test on the mechanical behavior of these coatings. The p-h plot of AlCrN film presented in Figure 11.4(a) showed two pop-in points at 381 μN and 738 μN respectively. The first pop-in point could be attributed to penetration into the amorphous-adhesive transfer layer, whereas the second pop-in point could be the onset of plastic deformation in the CrN interlayer.

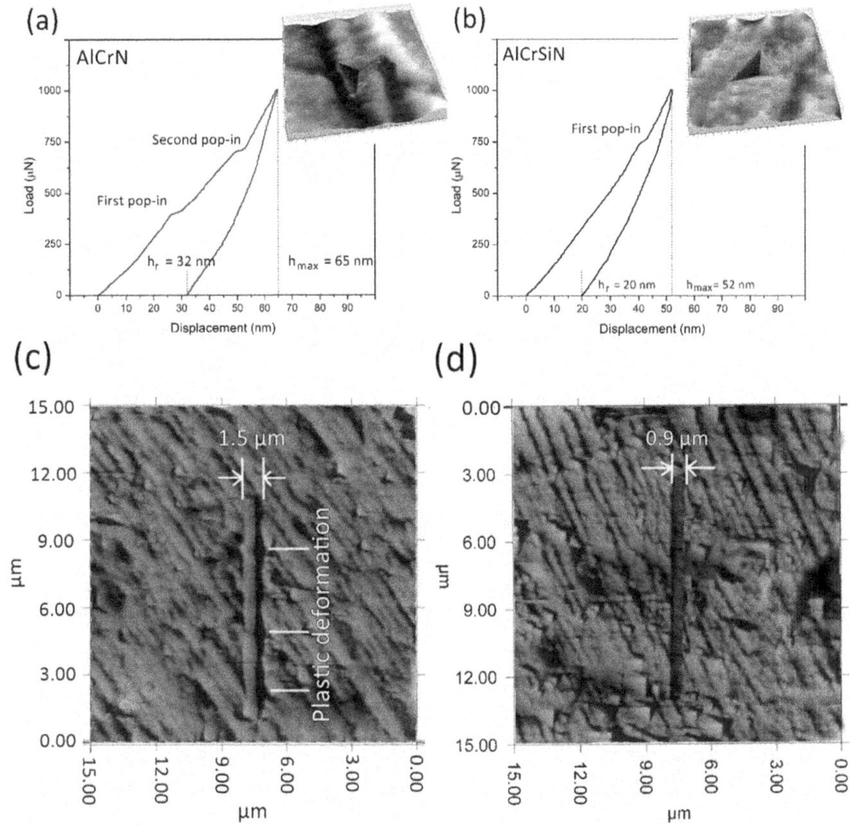

FIGURE 11.4 P-h plot of nanoindentation (with indentation dent insert) and SPM of nanoscratch test on worn surfaces of (a-b) AlCrN and (c-d) AlCrSiN.

On the other hand, the p-h plot shown in Figure 11.4(b) for the worn AlCrSiN film demonstrates a single pop-in point at 751 μN higher than both the pop-in points in worn AlCrN film. Here, it is important to note that AlCrSiN had registered lower h_{max} than AlCrN, leading to better retainment of surface hardness. The results of the nanoindentation test are tabulated in Table 11.2. Both the films have registered decrement in the mechanical properties after lubricated sliding. However, in case of AlCrN, a statistically significant difference ($p<0.05$) was seen between the nanoindentation and nanoscratch particulars obtained with as-deposited coating and worn-out surface (see Table 11.2), whereas no statistically significant difference was found in case of as-deposited and worn-out AlCrSiN surface. Owing to significant retainment of H/E and H^3/E^2, the worn AlCrSiN coating registered lower width of nanoscratch wear track (see Figure 11.4(d)). Whereas in the case of AlCrN, the sliding induced increased plastic work with significant loss in H/E and H^3/E^2 in the film, which is evident from the presence of plastic deformation at the edge of the scratch wear track (see Figure 11.4(c)).

TABLE 11.2
Nanoindentation and nanoscratch test results of the worn surfaces

Tests	Particulars	Thin Films		*p*-Value	Tukey's Grouping
		AlCrN	AlCrSiN		
Nanoindentation	Hardness (H in GPa)	21.3± 3.1	33.4±3.2	< 0.5	D
	Elastic modulus (E in GPa)	249±21	335±21	< 0.5	D
	Resistance to wear (H/E)	0.082 ±0.006	0.099 ± 0.006	< 0.5	D
	Resistance to deformation (H^3/E^2 in GPa)	0.15±0.11	0.33±0.14	> 0.5	S
	Elastic work (E.W. in pJ)	24 ± 5	40±6	< 0.5	D
	Plastic work (P.W, in pJ)	32±6	21±5	< 0.5	D
	Compressive stress (σ in GPa)	−1.8 ±0.6	3.3±0.6	>0.5	S
Nanoscratch	Scratch hardness (H_s in GPa)	24.3±3.4	35.5±4.7	< 0.5	D
	Friction coefficient	0.24±0.08	0.32±0.07	> 0.5	S

In line with this information, another aspect that can be considered for evaluating the resistance to wear is h_{max}/t factor [32]. A smaller h_{max}/t factor value for AlCrSiN than AlCrN is indicative of improved resistance to crack initiation and propagation [32]. What is more, previous studies [33, 34] suggest that elastic modulus ratio ($E_{coating}/E_{substrate}$) value of >1.3 depicts improved resistance to crack initiation and propagation upon loading in films. Worn AlCrN film had registered lower $E_{coating}/E_{substrate}$ value (1.25) attributed to amorphous nature of the adhesive film formed during lubricated sliding, the $E_{coating}/E_{substrate}$ for worn AlCrSiN was 1.67 (> 1.3). Vettivel et al. [35] had performed dry sliding test (in different parametric conditions) on TiN, TiAlN, and AlCrN films against Ti-alloy and observed high wear in AlCrN due to higher adhesion-induced wear. In another study, Mo et al. [23] had conducted dry reciprocating test on CrN and AlCrN film against pure titanium and reported inferior anti-adhesion properties of AlCrN-pure titanium tribopair.

The antibacterial performance of the as-deposited (control) and worn films against gram- negative *(Escherichia coli)* and gram-positive *(Bacillus subtilis)* bacteria was evaluated using surface spread plate method. The viable bacterial cell count (in CFU/mL) indicating the growth of gram-negative and gram-positive (see Figure 11.5(a)) bacterial colonies on the as-deposited and worn surfaces of the coatings indicate increase in the bacterial adhesion on the worn surfaces attributed to their inferior surface quality [14]. The SEM micrograph of the bacterial growth on the as-deposited surfaces is shown in Figure 11.5(b). Surface roughness plays an important role in bacterial adhesion [15]; increase in surface irregularities increases the effective surface area of contact, and cracks facilitates gram-positive bacteria adhesion. Further, surfaces demonstrating scratches, pits, and holes are prone to bacterial adhesion [14]. On the other hand, gram-negative bacterial adhesion is facilitated with increase in surface irregularities and asperities owing to the

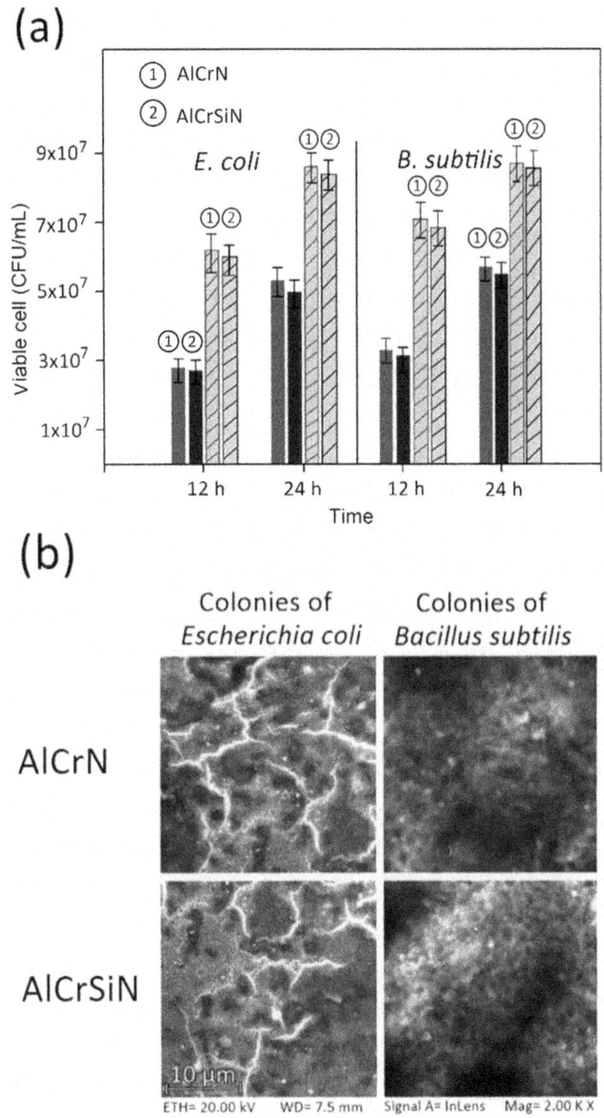

FIGURE 11.5 (a) Bacterial viable cell growth in CFU/mL on as-deposited and worn coated surfaces (b) SEM micrographs of the *E.coli* and *B. subtilis* colonies on the as-deposited surface.

presence of an extra membrane around their peptidoglycan layer, which aids in the surface interaction compared to rigid peptidoglycan layer in gram-positive bacteria [21, 15]. Further, the worn AlCrN surface demonstrated the formation of adhesive transfer layer from Ti6Al4V counterbody and possibly formed an amorphous TiO_2 layer. Almaguer-Flores et al. [16] had reported reduced bacterial adhesion in amorphous TiO_2 compared to crystalline structure due to the acidic nature of the latter which attracts bacterial adhesion. The antibacterial efficacy of the TiO_2 is further elucidated by Delgado et al. [36]. On the other hand, in the case of AlCrSiN, the hydrophilic and net positive charge of Si3N4 facilitates reduction in bacterial growth [18]. Further, stainless steel has reported to have showcased better bacterial anti-adhesion properties over titanium [37], signifying the suitability of stainless steel 316 LVM substrate chosen in the present study from antibacterial performance point of view. The antibacterial inhibition rate decreased from as-deposited to worn surface. For gram-positive bacteria, it had decreased by ~52% in both AlCrN and AlCrSiN films, whereas for gram-negative bacteria, it had decreased by ~56% in both the films. The CFUs obtained in the present study was comparable with the study of Hussein et al. [21], where they had obtained around ~7×10^7 CFU/ml and ~6.5×10^7 CFU/ml of gram-positive and gram-negative respectively after 24 h using surface spread plate method for as-deposited TiN film. Taken together, owing to degradation in surface quality in AlCrN and AlCrSiN films post lubricated tribotest, the bacterial adhesion on the worn coated surfaces had increased. However, the formation of amorphous TiO_2 on the AlCrN worn surface and the presence of amorphous Si_3N_4 in AlCrSiN could have negated the otherwise severe bacterial adhesion on the worn coated surface. Nonetheless, previous studies have reported the importance of studies on the bacterial adhesion and its growth in blood plasma medium [38, 39]. Thus, follow-up studies pertaining the bacterial adhesion in other relevant mediums can further corroborate the antibacterial properties of AlCrN and AlCrSiN films.

11.4 CONCLUSIONS

Si-doped AlCrN films, with a CrN interlayer, were deposited onto medical-grade 316 LVM stainless steel via cathodic-arc evaporation. Lubricated sliding tests simulating hip-implant conditions were then performed, and the following conclusions can be drawn:

a) In lubricated sliding against Ti6Al4V counterbody, AlCrN film exhibited higher friction coefficient due to formation and erosion of adhesive patch at the sliding interface resulting in plowing and galling on the worn coated surface. On the other hand, AlCrSiN coated surface witnessed fatigue-induced microcracks over it. Nanoindentation and nanoscratch test on the worn coated surfaces demonstrated better retainment of nonmechanical properties and structural integrity in worn AlCrSiN film compared to worn AlCrN due to the formation of an amorphous adhesive layer on the latter which is corroborated using Raman studies as well.
b) The biological response of unworn (control) and worn coated surfaces indicated slightly lower growth of *Escherichia coli* and *Bacillus subtilis* colonies on AlCrSiN film, possibly due to lower surface energy resulting in lower adhesion on it. However, no statistically significant difference was observed in the bacterial growth between both the films. The bacterial growth increased on the worn coated surfaces due to severe reduction in the surface quality of films post lubricated tribotest.

ACKNOWLEDGEMENT

The authors would like to thank Ingenious Tools and Design Solution Pvt. Ltd. Ghaziabad for providing assistance with thin film deposition and Central Research Facility, Indian Institute of Technology Kharagpur, Centre for Nanoscience and Nanotechnology, Sathyabama Institute of Science and Technology, Chennai, for conducting the experiments pertaining to the characterization of the films.

REFERENCES

1. Gao, Ying, Fei Cai, Xin Lu, Wei Xu, Ce Zhang, Jiazhen Zhang, and Xuanhui Qu. "Design of cycle structure on microstructure, mechanical properties and tribology behavior of AlCrN/AlCrSiN coatings." *Ceramics International* 48, no. 9 (2022): 12255–12270.
2. Nohava, Jiri, Pascal Dessarzin, Pavla Karvankova, and Marcus Morstein. "Characterization of tribological behavior and wear mechanisms of novel oxynitride PVD coatings designed for applications at high temperatures." *Tribology International* 81 (2015): 231–239.
3. Souza, Poliana S., Anderson J. Santos, Monique A. P. Cotrim, Alexandre M. Abrão, and Marcelo A. Câmara. "Analysis of the surface energy interactions in the tribological behavior of ALCrN and TIAlN coatings." *Tribology International* 146 (2020): 106206.
4. Li, Zhao, Paul Munroe, Zhong-Tao Jiang, Xiaoli Zhao, Jiang Xu, Zhi-Feng Zhou, Jian-Qing Jiang, Feng Fang, and Zong-Han Xie. "Designing superhard, self-toughening CrAlN coatings through grain boundary engineering." *Acta Materialia* 60, no. 16 (2012): 5735–5744.
5. Soldán, J., Jörg Neidhardt, B. Sartory, R. Kaindl, R. Čerstvý, Paul Heinz Mayrhofer, R. Tessadri, P. Polcik, M. Lechthaler, and Christian Mitterer. "Structure-property relations of arc-evaporated Al-Cr-Si-N coatings." *Surface and Coatings Technology* 202, no. 15 (2008): 3555–3562.
6. Warcholinski, Bogdan, Adam Gilewicz, Piotr Myslinski, Ewa Dobruchowska, Dawid Murzynski, and Tatyana A. Kuznetsova. "Effect of silicon concentration on the properties of Al-Cr-Si-N coatings deposited using cathodic arc evaporation." *Materials* 13, no. 21 (2020): 4717.
7. Patnaik, Lokeswar, Saikat Ranjan Maity, and Sunil Kumar. "Lubricated sliding of CFRPEEK/AlCrN film tribo-pair and its effect on the mechanical properties and structural integrity of the AlCrN film." *Materials Chemistry and Physics* 273 (2021): 124980.
8. Lillo, Cristian R., María Natalia Calienni, Roxana M. Gorojod, Maria Belen Rivas Aiello, Damián Rodriguez Sartori, María Jimena Prieto, Silvia del V. Alonso, Mónica L. Kotler, Mónica C. Gonzalez, and Jorge Montanari. "Toward biomedical application of amino-functionalized silicon nanoparticles." *Nanomedicine* 13, no. 11 (2018): 1349–1370.
9. Chang, Yin-Yu, Shi-Yao Weng, Chun-Hsiao Chen, and Fu-Xing Fu. "High temperature oxidation and cutting performance of AlCrN, TiVN and multilayered AlCrN/TiVN hard coatings." *Surface and Coatings Technology* 332 (2017): 494–503.
10. Polcar, Tomas, and Albano Cavaleiro. "High-temperature tribological properties of CrAlN, CrAlSiN and AlCrSiN coatings." *Surface and Coatings Technology* 206, no. 6 (2011): 1244–1251.
11. Patnaik, Lokeswar, Saikat Ranjan Maity, and Sunil Kumar. "Mechanical and tribological assessment of composite AlCrN or aC: Ag-based thin films for implant application." *Ceramics International* 47, no. 5 (2021): 6736–6752.
12. Basiaga, Marcin, Witold Walke, Wojciech Kajzer, Agata Sambok-Kiełbowicz, Janusz Szewczenko, Wojciech Simka, Marek Szindler, Bogusław Ziębowicz, and Vira Lubenets.

"Atomic layer deposited ZnO films on stainless steel for biomedical applications." *Archives of Civil and Mechanical Engineering* 21 (2021): 1–15.
13. Baad-Hansen, Thomas, Søren Kold, Poul Torben Nielsen, Mogens Berg Laursen, Poul Hedevang Christensen, and Kjeld Soballe. "Comparison of trabecular metal cups and titanium fiber-mesh cups in primary hip arthroplasty: A randomized RSA and bone mineral densitometry study of 50 hips." *Acta Orthopaedica* 82, no. 2 (2011): 155–160.
14. Bohinc, Klemen, Goran Dražić, Anže Abram, Mojca Jevšnik, Barbara Jeršek, Damijan Nipič, Marija Kurinčič, and Peter Raspor. "Metal surface characteristics dictate bacterial adhesion capacity." *International Journal of Adhesion and Adhesives* 68 (2016): 39–46.
15. Bagherifard, Sara, Daniel J. Hickey, Alba C. de Luca, Vera N. Malheiro, Athina E. Markaki, Mario Guagliano, and Thomas J. Webster. "The influence of nanostructured features on bacterial adhesion and bone cell functions on severely shot peened 316L stainless steel." *Biomaterials* 73 (2015): 185–197.
16. Almaguer-Flores, Argelia, Phaedra Silva-Bermudez, Rey Galicia, and Sandra E. Rodil. "Bacterial adhesion on amorphous and crystalline metal oxide coatings." *Materials Science and Engineering: C* 57 (2015): 88–99.
17. Sribalaji, M., O. S. Asiq Rahman, Tapas Laha, and Anup Kumar Keshri. "Nanoindentation and nanoscratch behavior of electroless deposited nickel-phosphorous coating." *Materials Chemistry and Physics* 177 (2016): 220–228.
18. Gorth, Deborah J., Sabrina Puckett, Batur Ercan, Thomas J. Webster, Mohamed Rahaman, and B. Sonny Bal. "Decreased bacteria activity on Si3N4 surfaces compared with PEEK or titanium." *International Journal of Nanomedicine* (2012): 4829–4840.
19. Sasikumar, Y., M. M. Solomon, L. O. Olasunkanmi, and E. E. Ebenso. "Effect of surface treatment on the bioactivity and electrochemical behavior of magnesium alloys in simulated body fluid." *Materials and Corrosion* 68, no. 7 (2017): 776–790.
20. Osés, J., J. F. Palacio, S. Kulkarni, A. Medrano, J. A. García, and R. Rodríguez. "Antibacterial PVD coatings doped with silver by ion implantation." *Applied Surface Science* 310 (2014): 56–61.
21. Hussein, Mohamed Abdrabou, N. K. Ankah, A. Madhan Kumar, Mohammed Abdol Azeem, S. Saravanan, Ahmad A. Sorour, and Nasser Al Aqeeli. "Mechanical, biocorrosion, and antibacterial properties of nanocrystalline TiN coating for orthopedic applications." *Ceramics International* 46, no. 11 (2020): 18573–18583.
22. Lee, R. Y., and Z. Eliezer. "Silicon segregation in sliding wear." *Wear* 87, no. 2 (1983): 227–233.
23. Mo, J. L., and M. H. Zhu. "Sliding tribological behaviors of PVD CrN and AlCrN coatings against Si3N4 ceramic and pure titanium." *Wear* 267, no. 5–8 (2009): 874–881.
24. Chen, Wanglin, Daoda Zhang, Dengcan Yao, Shihong Zhang, and Weiwei Wu. "Investigations on microstructure and mechanical properties of containing-Si coatings." *Surface Engineering* 33, no. 7 (2017): 536–541.
25. Xin, Zhiyuan, Xiaoran Yin, Yunhan Ling, Zhengjun Zhang, Xiao Liu, Hao Liang, and Xiukun Deng. "Hydrogen permeation behavior and mechanism of Cr2O3/Al2O3 bipolar oxide film under plasma discharging environment." *International Journal of Hydrogen Energy* 42, no. 32 (2017): 20869–20878.
26. Larbi, T., B. Ouni, A. Gantassi, K. Doll, M. Amlouk, and T. Manoubi. "Structural, optical and vibrational properties of Cr2O3 with ferromagnetic and antiferromagnetic order: A combined experimental and density functional theory study." *Journal of Magnetism and Magnetic Materials* 444 (2017): 16–22.
27. Tian, Hanjing, Elizabeth I. Ross, and Israel E. Wachs. "Quantitative determination of the speciation of surface vanadium oxides and their catalytic activity." *The Journal of Physical Chemistry B* 110, no. 19 (2006): 9593–9600.
28. Ma, Kai, Rui Zhang, Junlong Sun, and Changxia Liu. "Oxidation mechanism of biomedical titanium alloy surface and experiment." *International Journal of Corrosion* 2020, no. 1 (2020).

29. Smeltzer, W. W. "Oxidation of aluminum in the temperature range 400–600 C." *Journal of the Electrochemical Society* 103, no. 4 (1956): 209.
30. Wilson, J. R., and M. E. Lewis. "Oxidation of vanadium in dry and moist oxygen-argon mixtures." *Nature* 206, no. 4991 (1965): 1350–1351.
31. Mouayd, Amine Alaoui, Alexey Koltsov, Eliane Sutter, and Bernard Tribollet. "Effect of silicon content in steel and oxidation temperature on scale growth and morphology." *Materials Chemistry and Physics* 143, no. 3 (2014): 996–1004.
32. Patnaik, Lokeswar, Saikat Ranjan Maity, and Sunil Kumar. "Evaluation of crack resistance and adhesive energy of AlCrN and Ag doped aC films deposited on chrome nitrided 316 LVM stainless steel." *Advances in Materials and Processing Technologies* 8, no. sup3 (2022): 1048–1069.
33. Meyers, Marc A., A. Mishra, and David J. Benson. "Mechanical properties of nanocrystalline materials." *Progress in Materials Science* 51, no. 4 (2006): 427–556.
34. Łępicka, Magdalena, Małgorzata Grądzka-Dahlke, Daniel Pieniak, Kamil Pasierbiewicz, Kamila Kryńska, and Andrzej Niewczas. "Tribological performance of titanium nitride coatings: A comparative study on TiN-coated stainless steel and titanium alloy." *Wear* 422 (2019): 68–80.
35. Vettivel, S. C., R. Jegan, J. Vignesh, and S. Suresh. "Surface characteristics and wear depth profile of the TiN, TiAlN and AlCrN coated stainless steel in dry sliding wear condition." *Surfaces and Interfaces* 6 (2017): 1–10.
36. Delgado, L. P., M. Z. Figueroa-Torres, M. C. Ceballos-Chuc, R. García-Rodríguez, J. J. Alvarado-Gil, G. Oskam, and G. Rodriguez-Gattorno. "Tailoring the TiO2 phases through microwave-assisted hydrothermal synthesis: Comparative assessment of bactericidal activity." *Materials Science and Engineering: C* 117 (2020): 111290.
37. Malhotra, Rajesh, Benu Dhawan, Bhavuk Garg, Vivek Shankar, and Tapas Chandra Nag. "A comparison of bacterial adhesion and biofilm formation on commonly used orthopaedic metal implant materials: An in vitro study." *Indian Journal of Orthopaedics* 53 (2019): 148–153.
38. Ishikawa, Masahiro, Karen L. de Mesy Bentley, Bryan J. McEntire, B. Sonny Bal, Edward M. Schwarz, and Chao Xie. "Surface topography of silicon nitride affects antimicrobial and osseointegrative properties of tibial implants in a murine model." *Journal of Biomedical Materials Research Part A* 105, no. 12 (2017): 3413–3421.
39. Gunaratnam, Gubesh, Christian Spengler, Simone Trautmann, Philipp Jung, Johannes Mischo, Ben Wieland, Carlos Metz, Sören L. Becker, Matthias Hannig, Karin Jacobs, and Markus Bischoff. "Human blood plasma factors affect the adhesion kinetics of Staphylococcus aureus to central venous catheters." *Scientific Reports* 10, no. 1 (2020): 20992.

12 Multi-Scale Signal and Machine Learning-Driven Classification of Wear Regimes in Si-DLC/CFR-PEEK Tribosystems Under Dry Sliding Conditions

Jatin Gajjar and Amartya Shekhar Das

12.1 INTRODUCTION

Tribology plays a crucial role in determining the operational performance, energy efficiency, and life expectancy of engineered systems. The surfaces of components that work under high loads, dry contact, or back-and-forth motion often face mechanical and thermal stresses. These stresses can cause wear-related failures. Such failures are especially common in high-precision systems, including aerospace mechanisms, biomedical implants, and micro-electromechanical systems (MEMS), where keeping the surfaces intact is essential [1]. As the demand for smaller, more reliable systems that require little maintenance grows, surface engineering solutions, particularly through coatings, have received significant attention. Diamond-like carbon (DLC) coatings are among the most effective thin films for improving the surface properties of mechanical parts. Their high hardness, low coefficient of friction (COF), and chemical resistance make them ideal for many tribological applications [2]. Among the various modifications of DLC, doping with silicon (Si) has shown to markedly improve thermal stability, reduce residual stress, and promote the formation of self-lubricating tribofilms under dry contact [3, 4]. These tribofilms are particularly effective in reducing interfacial shear, thereby minimising wear and extending service life. Silicon-doped DLC (Si-DLC) coatings, therefore, have become the material of choice in many high-performance sectors. Parallel to the advancement of coating materials, polymeric counterbodies are increasingly being integrated into load-bearing systems, especially in medical and aerospace industries. One such material, carbon fibre-reinforced polyether ether ketone (CFR-PEEK),

DOI: 10.1201/9781003750659-12

offers a unique combination of low density, high fatigue strength, and excellent chemical resistance [5]. CFR-PEEK has emerged as a preferred polymer for articulating surfaces in orthopaedic implants and high-temperature bearings in aerospace components. However, tribological interactions between hard coatings like Si-DLC and soft, fibre-reinforced polymers such as CFR-PEEK remain relatively underexplored, particularly under dry sliding conditions where lubrication is absent and surface degradation is governed by complex physical and chemical phenomena [6]. Conventional tribological assessments typically rely on time-averaged metrics such as the mean COF, specific wear rate, or wear volume to evaluate performance. While these measures provide useful end-point data, they often mask transient instabilities, micro-events, or emergent frictional behaviours that occur during the sliding process. Consequently, the early detection of coating failure or regime transitions remains a challenging task for predictive maintenance [7].

To address this, in recent times, signal processing techniques are incorporated to capture the wear regimes and dynamics.

Among these, the continuous wavelet transform (CWT) has emerged as a comprehensive tool for analysing COF data. Unlike Fourier transforms, which operate solely in the frequency domain, CWT enables simultaneous time–frequency localisation, making it highly suitable for identifying intermittent and transient phenomena that reflect changes in wear mechanisms [8, 9]. Scalograms obtained from CWT allows understanding of energy evolution across different frequency bands, enabling the detection of transitions happening in wear regimes. CWT techniques has been in use for tribological systems that include rolling bearings [10], gear contacts [11], and biomedical tribo-interfaces [12]. In recent times, researchers have developed some other techniques such as return maps, which present the deterministic structure of a time series, and the scale index, a scalar metric derived from wavelet energy distribution that quantifies the degree of non-periodicity [13].

With the advent of Industry 4.0 and the proliferation of sensor-enabled systems, there is growing demand for integrating such signal analytics with machine learning (ML) techniques that can automate pattern recognition and support intelligent diagnostics [14]. ML algorithms, when trained on high-quality time-series data, can classify wear regimes, detect anomalies, and even predict time-to-failure without human intervention. Several recent studies have explored the application of ML in tribology. For instance, classification algorithms such as support vector machines (SVM), decision trees, and neural networks have been employed to detect bearing faults, predict lubricant degradation, and assess coating conditions [15, 16]. However, a few studies have attempted to combine multi-resolution signal analysis (e.g., CWT) with ML classification frameworks to provide a holistic view of dynamic wear processes. Even fewer have explored this integration in the context of Si-DLC coatings interfacing with polymeric materials like CFR-PEEK.

This study aims to bridge these gaps by presenting a hybrid analytical framework that integrates wavelet-based signal processing and supervised machine learning for wear regime classification in Si-DLC/CFR-PEEK tribosystems. Using a custom ball-on-disc tribometer, the COF signal was recorded at high resolution under dry sliding conditions. The signal was then segmented and analysed using CWT, return maps, and scale index analysis to visualise and quantify regime transitions. In parallel, an

ML model – specifically a Random Forest classifier – was trained on time-windowed signal features, including statistical descriptors (mean, RMS, skewness), spectral features (dominant frequency, spectral entropy), and complexity measures (permutation entropy, scale index). The goal was to classify the friction signal into three primary wear regimes: running-in, steady-state, and instable wear.

The novelty of this work lies in its comprehensive, data-driven methodology that fuses physical insight with computational intelligence. First, it is among the first studies to investigate the dynamic tribological behaviour of the Si-DLC/CFR-PEEK system using both wavelet analysis and machine learning. Second, it introduces a real-time capable wear regime classification system based on signal-derived features, moving beyond conventional tribological metrics. Third, it demonstrates how entropy and non-linearity indicators can serve not just as post-processing tools but as active features in automated classification models. The broader implication of this research is the development of intelligent monitoring tools that can be embedded into tribological test rigs or field-deployed components. Such tools could enable condition-based maintenance, extend component life, and reduce downtime in critical systems. For applications in orthopaedic implants, the ability to monitor wear in real time could lead to earlier detection of implant failure and more personalised patient care.

In conclusion, this study represents a significant step towards the digitalisation of tribological diagnostics, offering a framework that is both interpretive and predictive. By combining wavelet-based physical analysis with the classification power of machine learning, we demonstrate a pathway for building next-generation smart tribosystems capable of autonomous health monitoring.

12.2 EXPERIMENTAL SET-UP

In this study, a tribological system comprising a silicon-doped diamond-like carbon (Si-DLC) -coated stainless steel disc and a carbon fibre-reinforced polyether ether ketone (CFR-PEEK) counterbody was investigated under dry sliding conditions. The disc substrates were machined from cold-rolled AISI 316L stainless steel to a diameter of 25 mm and a thickness of 6 mm. Surface finishing was accomplished through successive stages of mechanical polishing using diamond suspensions down to 0.05 μm, achieving a mirror-like surface with an average roughness (Ra) of approximately 0.02 μm.

Si-DLC coatings were deposited via plasma-enhanced chemical vapour deposition (PECVD) using methane and tetramethylsilane (TMS) precursors. The coating process was conducted at 300 °C in a low-pressure RF plasma reactor, operating at a base pressure of 10^{-3} Pa and a power of 200 W. The resulting coatings exhibited a homogeneous hydrogenated amorphous carbon matrix, with a thickness of 2 μm as measured by contact profilometry. SEM characterisation revealed uniform nodular surface morphology typical of DLC coatings, and EDS analysis confirmed the incorporation of ~6 atomic percent silicon [17]. The counterbody was a 6 mm spherical pin fabricated from high-performance CFR-PEEK, a thermoplastic composite selected for its dimensional stability and favourable wear compatibility with carbon-based coatings [18]. Before testing, all samples were ultrasonically cleaned in acetone, isopropyl alcohol, and deionised water for 10 minutes each to eliminate contaminants and ensure consistent surface conditions [19]. Dry sliding tribological tests were performed using a ball-on-disc microtribometer at ambient temperature (25±2 °C) and 50% relative humidity. The disc was rotated at a constant linear sliding velocity of 0.10 m/s along a 6 mm wear track radius, while a 10 N normal load was applied vertically to the stationary CFR-PEEK ball. The total test duration was 60 minutes, and the coefficient of friction (COF) was sampled continuously at 20 Hz, resulting in a dataset of 72,000 points. For post-experiment analysis, wear track geometry was examined using contact profilometry and scanning electron microscopy. Based on visual and signal-based observations, the experiment was conceptually segmented into three wear regimes: running-in (0–10 min), steady state (10–40 min), and instable wear (40–60 min).

12.3 SIGNAL ANALYSIS USING WAVELET AND ENTROPY-BASED METHODS

To investigate the dynamic nature of the frictional system and to characterise regime transitions, the COF signal was analysed using a hybrid methodology combining continuous wavelet transform (CWT), return maps, scale index analysis (SIA), and time-localised permutation entropy (TPE). These tools provide complementary insights into time–frequency behaviour, non-linearity, and signal complexity.

12.3.1 Continuous Wavelet Transform (CWT)

The continuous wavelet transform (CWT) was applied to the COF signal using a complex Morlet wavelet with a central frequency of $\omega_0 = 6$. The CWT is defined as:

$$W(s,n) = \frac{1}{\sqrt{s}} \sum_{i=1}^{N} \mu(t_i) \cdot \psi^* \left(\frac{t_i - n}{s} \right)$$

where s is the scale (inversely proportional to frequency), n is the time translation, and $\psi^*(\cdot)$ denotes the complex conjugate of the wavelet function. The wavelet power spectrum (WPS), calculated as $P(s,n) = |W(s,n)|^2$, was used to visualise energy density over time and frequency [20].

Scalograms generated for each regime (Figure 12.1) revealed distinct time–frequency structures. The running-in phase exhibited burst-like low-frequency components due to initial asperity contact and mechanical adaptation. The steady-state phase displayed narrowband features suggestive of consistent sliding, while the instable wear phase showed broadband, chaotic energy spikes linked to coating degradation and debris entrainment [21].

12.3.2 Return Maps

Return maps were constructed by plotting $\mu(t+1)$ against $\mu(t)$ to investigate the deterministic structure of the signal. In the steady-state phase, the map displayed tight clustering, indicating quasi-periodic behaviour. In contrast, the running-in and instable wear phases exhibited greater scatter and asymmetry, suggesting increased stochasticity and potential chaotic dynamics (Figure 12.2) [22].

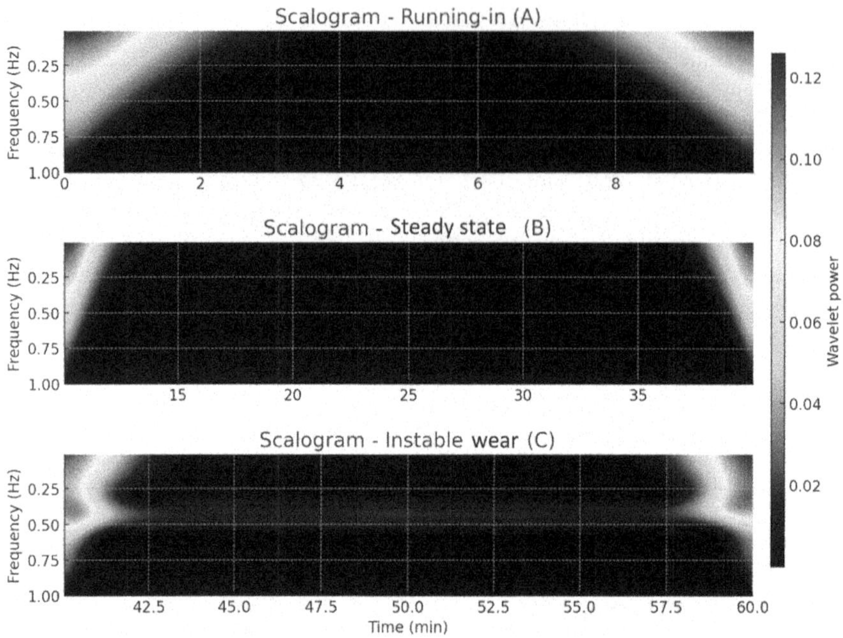

FIGURE 12.1 Wavelet scalograms of COF signal for different tribological states.

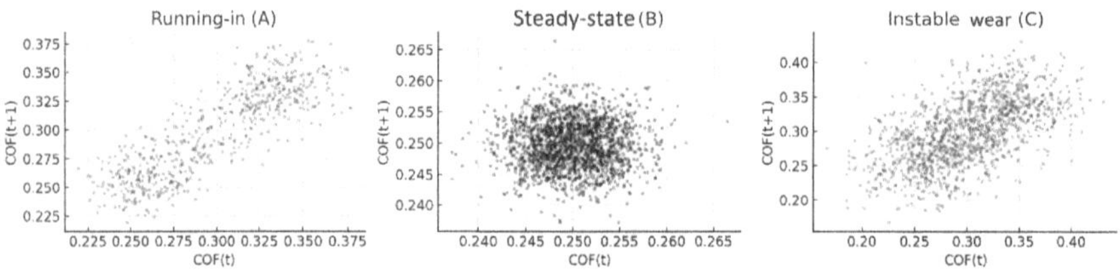

FIGURE 12.2 Return maps of COF signal for different tribological states.

12.3.3 Scale Index Analysis (SIA)

The scale index was used as a quantitative metric to assess the degree of periodicity within each segment of the COF signal. Based on the methodology proposed by Benitez et al. [23], the scale index i_{scale} is defined as:

$$i_{scale} = \frac{S(s_{min})}{S(s_{max})}$$

Here, $S(s)$ represents the time-averaged wavelet power at scale s, and the ratio reflects the deviation from periodicity. Lower values indicate more ordered behaviour [24]. The calculated scale indices were 0.176, 0.192, and 0.287 for the running-in, steady-state, and instable wear phases respectively (Figure 12.3), corroborating the observed trend of increasing non-linearity and aperiodicity with progressing wear.

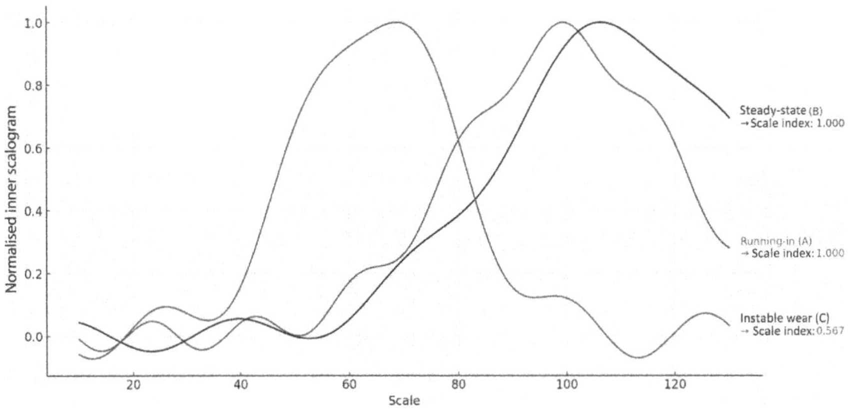

FIGURE 12.3 Inner scalograms and scale indices for tribological states.

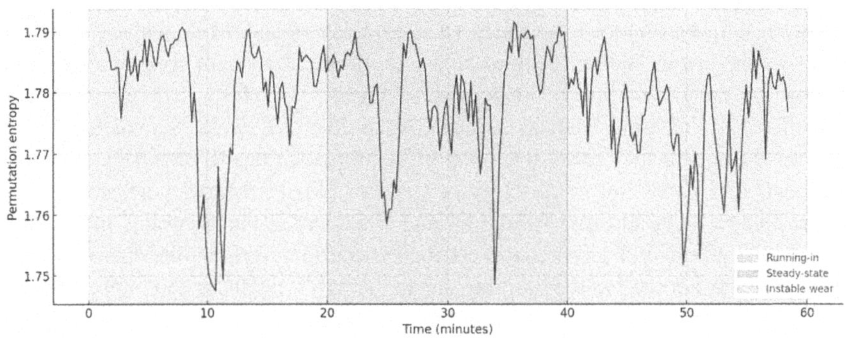

FIGURE 12.4 Time-localised permutation entropy of COF signal.

12.3.4 Time-Localised Permutation Entropy (TPE)

To evaluate the local complexity of the friction signal, time-localised permutation entropy was computed using sliding windows of 400 samples (20 s) with 100-sample strides (5 s). Permutation entropy, introduced as a model-free metric of complexity, is defined as:

$$H = -\sum_{i=1}^{k} p_i log_2 p_i$$

where p_i denotes the probability of the i-th ordinal pattern among $k=m!$ possible permutations. The entropy is normalised by $\log_2(k)$, ensuring $H \in [0,1]$. In this analysis, an embedding dimension $m = 3$ and time delay $\tau = 1$ were used [25].

The resulting entropy profile (Figure 12.4) demonstrated a moderate complexity during the running-in phase, a noticeable reduction in entropy during steady state (indicating stable frictional dynamics), and a significant rise in the instable wear phase, reflecting system destabilisation. These observations align with recent applications of permutation entropy in tribology and fault diagnostics, where entropy surges mark transitions to stochastic or chaotic behaviour.

12.3.5 Machine Learning-Driven Regime Classification Framework

To execute a robust evaluation of the tribological system, a supervised ML framework was implemented that has the potential to classify the wear regimes taking COF data into consideration. This technique enables real-time diagnostics and predictive maintenance by enabling autonomous identification of regime transitions from complex time-series data [26]. The COF signal, recorded at a sampling rate of 20 Hz over 60 minutes, was divided into a sliding window of 400 samples, or 20 seconds, with a 50% overlap. Each segment was labelled based on its timestamp: the first 10 minutes were marked as running-in, the next 10–40 minutes as steady-state, and the last 20 minutes as instable wear. These labelled segments provided data associated to the wear regimes. From each segment, a set of 15 statistical and dynamical features was extracted to capture the temporal, spectral, and complexity characteristics of the signal.

These features included measures such as mean, standard deviation, skewness, kurtosis, root mean square, peak-to-peak amplitude, dominant frequency, spectral entropy, permutation entropy, scale index, energy, and zero-crossing rate. After normalising the feature matrix, several ML algorithms were trained and evaluated, including logistic regression, support vector machines (SVM), random forest, XGBoost, and a shallow multi-layer perceptron (MLP). Model performance was assessed using fivefold cross-validation and evaluated using accuracy, macro-averaged precision, recall, F1-score, and confusion matrices. Among the classifiers, random forest was adopted as the baseline due to its high robustness and interpretability, with feature importance analysis offering insights into the dominant signal characteristics governing wear regime classification [27].

12.4 RESULTS AND DISCUSSION

Understanding the tribological performance of Si-DLC coatings against CFR-PEEK counterbodies necessitates a multipronged signal analysis framework. This study applied four complementary methods – continuous wavelet transform (CWT), return maps, inner scalogram-based scale index (SI), and time-localised permutation entropy (TPE) – to extract and interpret the underlying physical and dynamical mechanisms embedded in the COF signal across three well-defined wear regimes: running-in, steady-state, and unstable wear. This section presents the findings and inter-method consistencies revealed by these tools, along with contextualisation in existing literature and potential implications.

12.4.1 Time–Frequency Domain Characteristics: CWT Analysis

Figure 12.1 displays the wavelet scalograms for the three identified wear regimes using the complex Morlet wavelet. The scalograms visualise the distribution of spectral energy over time, allowing a direct interpretation of the evolution of tribological events. In the running-in phase, energy is concentrated between 4 Hz and 8 Hz, consistent with early surface conformity processes such as micro-asperity wear, oxide film formation, and debris entrapment [28]. These frequencies reflect transitional phenomena where frictional energy fluctuates due to variable micro-contact geometry. During the steady-state phase, the scalogram shows strong energy localisation below 3 Hz. This suggests minimal vibration and dynamic stability, indicative of a matured tribological interface where conformal contact, low wear, and stable COF values dominate [29]. Si-DLC coatings are known for maintaining low friction coefficients under steady load conditions due to their lubricious transfer film formation and resistance to oxidative degradation [30]. In contrast, the unstable wear phase exhibits a broadband frequency response between 6 Hz and 15 Hz, with recurrent high-energy bursts. These dynamics likely originate from micro-fracture events, loss of coating integrity, and abrupt transitions due to localised overheating or third-body wear [31]. The presence of frequency scattering in this phase demonstrates the system's transition to chaotic and energetically inefficient behaviour. The CWT analysis confirms that tribological transitions are reflected in distinct spectral fingerprints. Compared to conventional Fourier analysis, the CWT captures transient dynamics more accurately and supports early stage detection of unstable wear regimes [32].

12.4.2 Dynamical Determinism: Return Maps

Return maps (COF(t) versus COF(t+1)) provide insights into the underlying dynamical behaviour of the frictional process. As shown in Figure 12.2, the three regimes exhibit markedly different attractor structures. In the running-in regime, the return map shows an elongated cloud with moderate dispersion, signifying weak determinism. The mechanical interactions at this stage involve variable contact geometries, making the COF evolution sensitive to microscopic asperity breakdowns [33]. The steady-state regime displays the most deterministic pattern, with points clustered along the diagonal. This suggests strong correlation between consecutive COF values and stable phase space behaviour. Here, the tribological interface operates near a mechanical equilibrium with minimal variation [34].

In the unstable wear phase, the map becomes diffuse with no clear structure, indicating high unpredictability. This reflects a dynamic loss of memory in the system and the dominance of stochastic processes such as particle-induced vibration and thermal softening of the PEEK matrix [35]. Overall, return maps validate the temporal consistency of the tribosystem and can act as fast heuristics for identifying the onset of instability.

12.4.3 Multi-Scale Energy Distribution: Inner Scalogram and Scale Index

To quantify the energy distribution across frequency scales, we computed inner scalograms and derived the scale index (SI) for each regime. Figure 12.3 shows the normalised energy versus wavelet scale curves. The running-in regime exhibits moderate spectral spread, with energy concentrated in the mid-scale range (~20–40), signifying intermediate temporal structures [36]. The steady-state regime shows a narrow and steep energy curve, localised to low scales (< 15), indicating high periodicity and minimal multiscale disturbances. This aligns with the behaviour of well-functioning coatings that experience negligible surface disruption [37]. Conversely, the unstable wear phase demonstrates the broadest energy spread, with significant power in both low and high scales. This indicates the presence of multiple frequencies and irregular dynamics, corroborating the insights from the CWT and return maps.

The corresponding SI values – 0.176 (running-in), 0.192 (steady-state), and 0.287 (unstable) – capture the regime-wise progression in dynamic complexity. A higher SI signifies greater distribution of energy across scales, consistent with increased disorder and degradation. The inner scalogram analysis thus bridges the frequency and scale domains and provides a scalar metric to monitor tribological stability over time.

12.4.4 Temporal Complexity Evolution: Time-Localised Permutation Entropy

Figure 12.4 illustrates the permutation entropy (PE) profiles computed using a sliding window across the entire COF signal. During running-in, entropy fluctuates between 0.55 and 0.65, suggesting moderate variability in ordinal patterns. This reflects the transitional nature of this regime where deterministic and stochastic forces coexist [38].

The steady-state regime yields the lowest PE (~0.45), indicative of a stable and quasi-periodic signal. This aligns with the low SI and deterministic return map, suggesting low system complexity and consistent interfacial mechanics [37]. Entropy surges during the unstable wear phase, peaking around 0.85. Such high values correspond to significant dynamical disorder, resulting from wear particle interference and frictional heating, disrupting any latent periodicity [39]. PE complements the wavelet and scale index methods by offering a real-time complexity score that is robust to noise and sensitive to local transitions.

12.4.5 Machine Learning-Driven Regime Classification

To enable predictive diagnostics and automated detection of wear transitions, an ML approach was integrated into the tribological framework. A supervised classification model was developed using labelled features extracted from the COF signal. The goal was to classify the three distinct wear regimes – running-in, steady-state, and instable wear – based on their underlying dynamical signatures.

The COF signal, recorded at a resolution of 20 Hz over a 60-minute duration, was segmented into overlapping 20-second windows with a 50% overlap. Each segment was labelled according to its timestamp, and 15 features were extracted from each segment. These included statistical descriptors (e.g., mean, standard deviation), frequency-domain features (e.g., spectral entropy, dominant frequency), and non-linear complexity metrics (e.g., permutation entropy and scale index).

A random forest classifier was trained and validated using fivefold cross-validation. As shown in Figure 12.5, the model achieved an overall accuracy of over 92%, with strong precision and recall across all classes. The confusion matrix illustrates a high

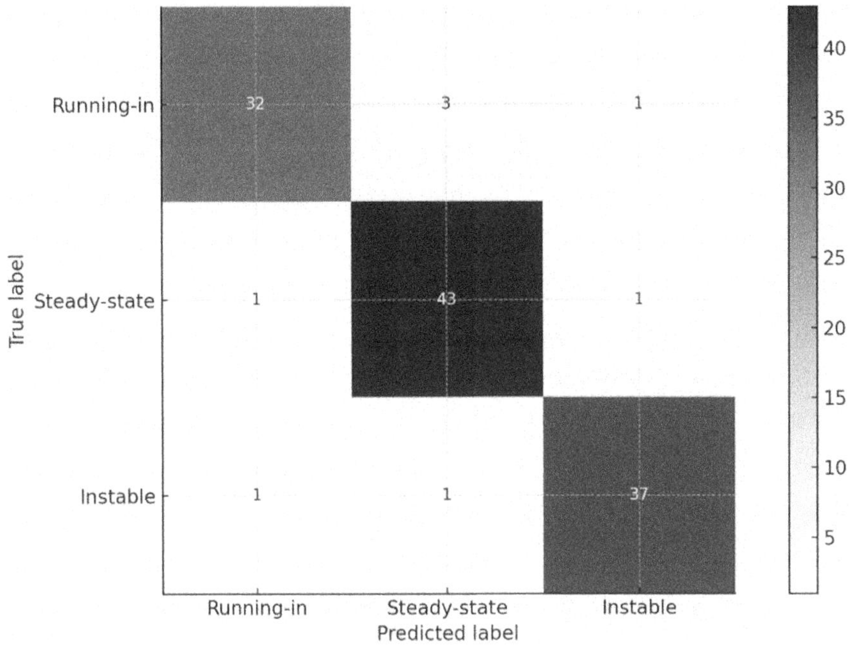

FIGURE 12.5 Confusion matrix for wear regime classification.

degree of discrimination between the wear regimes, especially between the steady-state and instable phases – an otherwise challenging boundary due to transitional fluctuations. Feature importance analysis revealed that non-linear and spectral complexity features – namely permutation entropy, scale index, spectral entropy, and dominant frequency – were the most influential predictors for classification as shown in Figure 12.6. This aligns with previous findings that tribological transitions are better characterised through multi-scale and complexity-based metrics rather than purely statistical summaries [40, 41]. The importance of entropy-based features supports the hypothesis that chaotic and stochastic frictional behaviour becomes more pronounced as the system enters the unstable regime.

The results suggest that the steady-state regime is defined by low entropy and low-scale index values, indicating stable and periodic contact mechanics. Conversely, the instable wear regime is characterised by high spectral dispersion and elevated entropy, marking the onset of system-level degradation, possibly due to delamination, third-body effects, or thermal softening of the CFR-PEEK matrix [42]. The use of machine learning in tribology is gaining traction for its ability to uncover latent patterns in frictional data and support condition-based monitoring [43]. The approach demonstrated here provides a practical pathway for integrating ML-based classifiers into real-time wear detection modules, particularly in biomedical implants and aerospace actuators, where system failure must be predicted and mitigated at early stages.

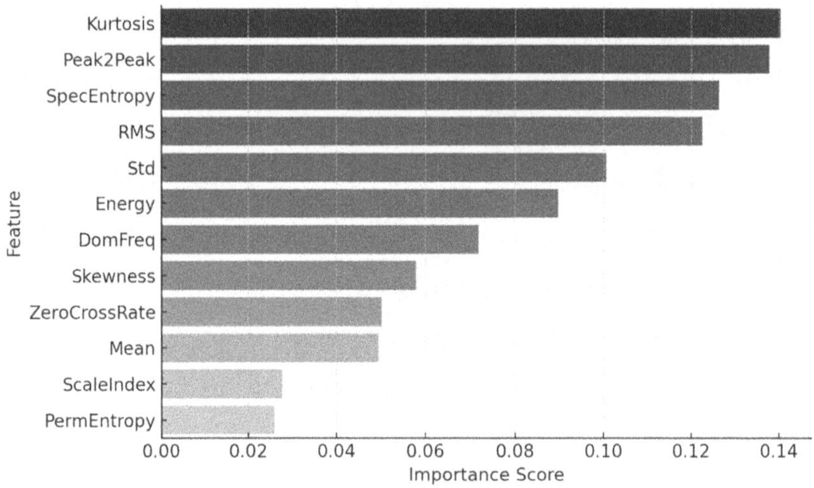

FIGURE 12.6 Feature importance from random forest classifier.

12.4.6 Cross-Validation

All four analytical approaches – continuous wavelet transform (CWT), return maps, inner scalograms with scale index (SI), and time-localised permutation entropy (TPE) – collectively unravel the evolving tribo-dynamic landscape of the Si-DLC/CFR-PEEK contact. These methods provide converging evidence for the existence of three tribological states and allow in-depth insight into the mechanical and dynamical transitions occurring during the wear process. The CWT reveals regime-dependent time–frequency patterns, with energy localisation observed during steady state and spectral broadening under instability. In parallel, return maps differentiate the dynamical determinism of each regime, showing structured trajectories during steady state and randomness under unstable wear. The inner scalogram and SI capture energy distributions over frequency scales, offering a scalar representation of multiscale spectral activity, while TPE quantifies complexity evolution over time by capturing ordinal pattern randomness within the COF signal. Notably, both SI and TPE show sensitivity to critical transitions, making them potential markers for early detection of wear-induced disorder.

This layered analytical framework goes beyond traditional metrics such as mean COF or RMS variation. For instance, the scale index increases from 0.176 to 0.287, while TPE rises from 0.45 in steady state to ~0.85 in the unstable phase. These changes independently signal an increase in dynamical disorder, suggesting convergence between time–frequency and complexity-based methods in identifying wear transitions [44, 45]. While some prior works have utilised entropy-based diagnostics or fractal analysis for condition monitoring [46], the combination of wavelet-based spectral measures with permutation entropy is novel. The inner scalogram, in particular, allows a compressed representation of time–frequency energy and, when paired with TPE, provides orthogonal insights – one in frequency-scale space, the other in temporal pattern complexity [47, 48]. Recent advancements in fault diagnosis also highlight the strength of hybrid complexity-wavelet pipelines in characterising non-stationary signals, lending credence to the relevance of our approach in tribology [49]. The findings also shed light on material-specific behaviour. Si-DLC coatings demonstrate excellent steady-state performance due to their intrinsic hardness and graphitic transfer layers as observed from the narrow frequency bands and low entropy values. However, once mechanical thresholds are exceeded – likely due to cumulative stress, thermal gradients, or counterface degradation – both SI and entropy metrics suggest rapid destabilisation. The CFR-PEEK counterface, with its polymeric nature, introduces inherent damping and debris formation, which further complicates the frictional interface in late wear phases [50]. From an application standpoint, such diagnostics could enable condition-based monitoring systems in medical implants or aerospace actuators, where surface damage must be detected non-invasively and in real time. Embedding SI and TPE analytics into sensor firmware could offer edge-diagnostics capabilities, reducing the dependency on offline analysis [51].

Finally, while each method stands independently, the synthesis of scale-aware and entropy-driven methods provides a richer understanding of friction-induced transitions. This cross-validated approach marks a methodological advance in tribological signal analysis and offers a blueprint for future integration with ML models to enhance predictive wear analytics [52, 53].

12.5 CONCLUSIONS

This study presented a comprehensive, multi-method analysis of the tribological performance of silicon-doped DLC-coated stainless steel sliding against a CFR-PEEK counterbody under dry conditions. By integrating five advanced signal processing techniques – CWT, return maps, delay coordinate mapping, inner scalogram with scale index, and time-localised permutation entropy – we were able to robustly segment and characterise the wear process into three regimes: running-in, steady-state, and unstable wear. The CWT scalograms illustrated how spectral energy distribution evolves with time, revealing regime transitions in a time–frequency context.

The return maps and delay plots showed reduced predictability with increased instability in the wear regimes. Inner scalograms showed shifts in multiscale energy concentrations which was further quantified using scale index. Further, TPE provided the dynamic measure of signal complexity by capturing early signs of change in disorder or change to unstable wear regime. To put things on perspective, Si-DLC showcased low friction properties during steady state phase; however, with the progression of wear, the wear interface became unstable owing to plausible wear debris, which was indicated by high entropy and dispersed spectral content.

REFERENCES

1. Jacques, Kelly, Nikhil Murthy, Satish Dixit, Diana Berman, and Stephen Berkebile. "Method for Tribological Experiment to Study Scuffing Initiation on AISI 52100 Steel and Hard Ceramic Coatings." *Tribology International* 160 (2021): 107001.

2. Marian, Max, and Stephan Tremmel. "Current Trends and Applications of Machine Learning in Tribology – A Review." *Lubricants* 9, no. 9 (2021): 86.
3. Rosenkranz, Andreas, Max Marian, Francisco J. Profito, Nathan Aragon, and Raj Shah. "The Use of Artificial Intelligence in Tribology – A Perspective." *Lubricants* 9, no. 1 (2021): 2.
4. Sose, Anusha T., Dharmendra Misra, Pradeep R. Guduru, and Amit K. Roy. "A Review of Recent Advances and Applications of Machine Learning in Tribology." *Physical Chemistry Chemical Physics* 25 (2023): 4408–4443.
5. Chen, Shengshan, Ganlin Cheng, Fei Guo, Xiaohong Jia, and Xiaohao Wen. "Integrating Friction Noise for In Situ Monitoring of Polymer Wear Performance: A Machine Learning Approach in Tribology." *Journal of Tribology* 147, no. 6 (2025): 061701.
6. Shandhoosh, V., Ganjikunta Chakrapani, V. Sugumaran, Sangharatna M. Ramteke, and Max Marian. "Intelligent Fault Diagnosis for Tribo-Mechanical Systems by Machine Learning: Multi-Feature Extraction and Ensemble Voting Methods." *Knowledge-Based Systems* 305 (2024): 112694.
7. Bandt, Christoph, and Bernd Pompe. "Permutation Entropy: A Natural Complexity Measure for Time Series." *Physical Review Letters* 88, no. 17 (2002): 174102.
8. Rosso, Osvaldo A., H. A. Larrondo, María Teresa Martin, A. Plastino, and Miguel A. Fuentes. "Distinguishing Noise from Chaos." *Physical Review Letters* 99, no. 15 (2007): 154102.
9. Zhao, Li-Ye, Lei Wang, and Ru-Qiang Yan. "Rolling Bearing Fault Diagnosis Based on Wavelet Packet Decomposition and Multi-Scale Permutation Entropy." *Entropy* 17, no. 9 (2015): 6447–6461.
10. Wu, Shuen-De, Po-Hung Wu, Chiu-Wen Wu, Jian-Jiun Ding, and Chun-Chieh Wang. "Bearing Fault Diagnosis Based on Multiscale Permutation Entropy and Support Vector Machine." *Entropy* 14, no. 8 (2012): 1343–1356.
11. Humeau-Heurtier, Anne. "The Multiscale Entropy Algorithm and Its Variants: A Review." *Entropy* 17, no. 5 (2015): 3110–3123.
12. Li, Zhe, Yahui Cui, Longlong Li, Runlin Chen, Liang Dong, and Juan Du. "Hierarchical Amplitude-Aware Permutation Entropy-Based Fault Feature Extraction Method for Rolling Bearings." *Entropy* 24, no. 3 (2022): 310.
13. Bolós, Vicente J., Rafael Benítez, and Román Ferrer. "A New Wavelet Tool to Quantify Non-Periodicity of Non-Stationary Economic Time Series." *Mathematics* 8, no. 5 (2020): 844.
14. Yan, Ruqiang, Robert X. Gao, and Xuefeng Chen. "Wavelets for Fault Diagnosis of Rotary Machines: A Review with Applications." *Signal Processing* 96 (2014): 1–15.
15. Shaikh, Shahsharif, Mohammadamin Sadeghi, Sandra Cruz, and Fábio Ferreira. "Recent Progress on the Tribology of Pure/Doped Diamond-Like Carbon Coatings and Ionic Liquids." *Coatings* 14, no. 1 (2024): 71.
16. Zanin, Massimiliano, Luciano Zunino, Osvaldo A. Rosso, and David Papo. "Permutation Entropy and Its Main Biomedical and Econophysics Applications: A Review." *Entropy* 14, no. 8 (2012): 1553–1577.
17. Patnaik, Lokeswar, Sunil Kumar, Jatin Gajjar, Amarapalli Deepak, Jayketh Naidana, V. S. S. Venkatesh, Magdalena Lepicka, Saikat Ranjan Maity, Syed Mahammad Shafi, and Shiela Chetri. "Carbon-Fibre-Reinforced-PEEK and Silicon Doped Amorphous Carbon as a Potential Tribopair for Implant Application." *Advances in Materials and Processing Technologies* 10, no. 4 (2024): 2761–2779.
18. Patnaik, Lokeswar, Sunil Kumar, Saikat Ranjan Maity, Magdalena Łępicka, and A. Anderson. "Fuzzy CODAS Based Analysis of Wear and Temperature Induced Responses of Si Doped aC Film and CFRPEEK Tribopair." *Silicon* 15, no. 6 (2023): 2829–2847.
19. Shevchik, Sergey A., Fatemeh Saeidi, Bastian Meylan, and Kilian Wasmer. "Prediction of Failure in Lubricated Surfaces Using Acoustic Time-Frequency Features and Random

Forest Algorithm." *IEEE Transactions on Industrial Informatics* 13, no. 4 (2016): 1541–1553.

20. Srivastava, Madhur, C. Lindsay Anderson, and Jack H. Freed. "A New Wavelet Denoising Method for Selecting Decomposition Levels and Noise Thresholds." *IEEE Access* 4 (2016): 3862–3877.
21. Zhao, Li-Ye, Lei Wang, and Ru-Qiang Yan. "Rolling Bearing Fault Diagnosis Based on Wavelet Packet Decomposition and Multi-Scale Permutation Entropy." *Entropy* 17, no. 9 (2015): 6447–6461.
22. Liu, Ningyu, and Huajiang Ouyang. "Friction-Induced Vibration Considering Multiple Types of Nonlinearities." *Nonlinear Dynamics* 102, no. 4 (2020): 2057–2075.
23. Benítez, Rafael, Vicente J. Bolós, and M. E. Ramírez. "A Wavelet-Based Tool for Studying Non-Periodicity." *Computers & Mathematics with Applications* 60, no. 3 (2010): 634–641.
24. Riedl, Maik, Andreas Müller, and Niels Wessel. "Practical Considerations of Permutation Entropy: A Tutorial Review." *The European Physical Journal Special Topics* 222, no. 2 (2013): 249–262.
25. Zheng, Jinde, Zhilin Dong, Haiyang Pan, Qing Ni, Tao Liu, and Jun Zhang. "Composite Multi-Scale Weighted Permutation Entropy and Extreme Learning Machine Based Intelligent Fault Diagnosis for Rolling Bearing." *Measurement* 143 (2019): 69–80.
26. Marian, Max, and Stephan Tremmel. "Physics-Informed Machine Learning – an Emerging Trend in Tribology." *Lubricants* 11, no. 11 (2023): 463.
27. König, F., C. Sous, A. Ouald Chaib, and G. Jacobs. "Machine Learning Based Anomaly Detection and Classification of Acoustic Emission Events for Wear Monitoring in Sliding Bearing Systems." *Tribology International* 155 (2021): 106811.
28. Keller, Karsten, Teresa Mangold, Inga Stolz, and Jenna Werner. "Permutation Entropy: New Ideas and Challenges." *Entropy* 19, no. 3 (2017): 134.
29. Patnaik, Lokeswar, Saikat Ranjan Maity, and Sunil Kumar. "Effect of Lubricated Sliding Wear Against CFRPEEK on the Nanomechanical Properties of Ag Alloyed Cr/DLC Thin Film." *Journal of the Mechanical Behavior of Biomedical Materials* 118 (2021): 104478.
30. Patnaik, Lokeswar, Saikat Ranjan Maity, and Sunil Kumar. "Comprehensive Structural, Nanomechanical and Tribological Evaluation of Silver Doped DLC Thin Film Coating with Chromium Interlayer (Ag-DLC/Cr) for Biomedical Application." *Ceramics International* 46, no. 14 (2020): 22805–22818.
31. Patnaik, Lokeswar, Saikat Ranjan Maity, and Sunil Kumar. "Lubricated Sliding of CFRPEEK/AlCrN Film Tribo-Pair and Its Effect on the Mechanical Properties and Structural Integrity of the AlCrN Film." *Materials Chemistry and Physics* 273 (2021): 124980.
32. Baccar, Dorra, and Dirk Söffker. "Wear Detection by Means of Wavelet-Based Acoustic Emission Analysis." *Mechanical Systems and Signal Processing* 60, no. 61 (2015): 198–207.
33. Gao, Kangping, Xinxin Xu, and Shengjie Jiao. "Measurement and Prediction of Wear Volume of the Tool in Nonlinear Degradation Process Based on Multi-Sensor Information Fusion." *Engineering Failure Analysis* 136 (2022): 106164.
34. Gholizadeh, S., and O. A. Samavati. "Structural Optimization by Wavelet Transforms and Neural Networks." *Applied Mathematical Modelling* 35, no. 2 (2011): 915–929.
35. Yan, Ruqiang, Robert X. Gao, and Xuefeng Chen. "Wavelets for Fault Diagnosis of Rotary Machines: A Review with Applications." *Signal Processing* 96 (2014): 1–15.
36. Raous, Michel, Serge Barbarin, and Didier Vola. "Numerical Characterization and Computation of Dynamic Instabilities for Frictional Contact Problems." In *Friction and Instabilities*, pp. 233–291. Vienna: Springer, 2002.
37. Allaire, Douglas, Qinxian He, John Deyst, and Karen Willcox. "An Information-Theoretic Metric of System Complexity with Application to Engineering System Design." *Journal of Mechanical Design* (2012): 100906.

38. Kilundu, Bovic, Pierre Dehombreux, and Xavier Chiementin. "Tool Wear Monitoring by Machine Learning Techniques and Singular Spectrum Analysis." *Mechanical Systems and Signal Processing* 25, no. 1 (2011): 400–415.
39. Li, Yongbo, Xianzhi Wang, Zhenbao Liu, Xihui Liang, and Shubin Si. "The Entropy Algorithm and Its Variants in the Fault Diagnosis of Rotating Machinery: A Review." *IEEE Access* 6 (2018): 66723–66741.
40. Shah, Raj, Nikhil Pai, Gavin Thomas, Swarn Jha, Vikram Mittal, Khosro Shirvni, and Hong Liang. "Machine Learning in Wear Prediction." *Journal of Tribology* 147, no. 4 (2025): 040801.
41. Yin, Nian, Zhiguo Xing, Ke He, and Zhinan Zhang. "Tribo-Informatics Approaches in Tribology Research: A Review." *Friction* 11, no. 1 (2023): 1–22.
42. Kurdi, Abdulaziz, and Li Chang. "Recent Advances in High Performance Polymers – Tribological Aspects." *Lubricants* 7, no. 1 (2018): 2.
43. Kotha Amarnath, Swathi, Vamsi Inturi, Sabareesh Geetha Rajasekharan, and Amrita Priyadarshini. "Combining Sensor Fusion and a Machine Learning Framework for Accurate Tool Wear Prediction During Machining." *Machines* 13, no. 2 (2025): 132.
44. Nosonovsky, Michael. "Entropy in Tribology: In the Search for Applications." *Entropy* 12, no. 6 (2010): 1345–1390.
45. Ma, Chenyang, Ke Feng, Xianzhi Wang, Zhiqiang Cai, and Yongbo Li. "Condition-Adaptive Permutation Entropy: A Novel Dynamic Complexity-Based Health Indicator for Bearing Health Monitoring." *IEEE Transactions on Reliability* 74, no. 1 (2024): 2394–2407.
46. Stachowiak, Gwidon, and Andrew W. Batchelor. *Engineering Tribology*, 4th ed. Elsevier, 2013.
47. Uyulan, Caglar, Türker Tekin Ergüzel, and Nevzat Tarhan. "Entropy-Based Feature Extraction Technique in Conjunction with Wavelet Packet Transform for Multi-Mental Task Classification." *Biomedical Engineering/Biomedizinische Technik* 64, no. 5 (2019): 529–542.
48. Cencini, M., Massimo Falcioni, E. Olbrich, H. Kantz, and Angelo Vulpiani. "Chaos or Noise: Difficulties of a Distinction." *Physical Review E* 62, no. 1 (2000): 427.
49. Huang, Dajian, Wen-An Zhang, Fanghong Guo, Weijiang Liu, and Xiaoming Shi. "Wavelet Packet Decomposition-Based Multiscale CNN for Fault Diagnosis of Wind Turbine Gearbox." *IEEE Transactions on Cybernetics* 53, no. 1 (2021): 443–453.
50. Alpkaya, Alican Tuncay, and Şenay Mihçin. "The Computational Approach to Predicting Wear: Comparison of Wear Performance of CFR-PEEK and XLPE Liners in Total Hip Replacement." *Tribology Transactions* 66, no. 1 (2023): 59–72.
51. Lu, Siliang, Jingfeng Lu, Kang An, Xiaoxian Wang, and Qingbo He. "Edge Computing on IoT for Machine Signal Processing and Fault Diagnosis: A Review." *IEEE Internet of Things Journal* 10, no. 13 (2023): 11093–11116.
52. Sivaraman, S., N. Radhika, and Muhammad Abubaker Khan. "Machine Learning-Driven Prediction of Wear Rate and Phase Formation in High Entropy Alloy Coatings for Enhanced Durability and Performance." *IEEE Access* 13, (2025): 33956–33975.
53. Zhou, Zilong, Yue Yu, Chaoyang Song, Zhen Liu, Manman Shi, and Jingxiang Zhang. "Multi-View Weighted Feature Fusion with Wavelet Transform and CNN for Enhanced CT Image Recognition." *Journal of Intelligent & Fuzzy Systems* 45, no. 6 (2023): 12167–12183.

Index

For Product Safety Concerns and Information please contact our EU representative GPSR@taylorandfrancis.com
Taylor & Francis Verlag GmbH, Kaufingerstraße 24, 80331 München, Germany

www.ingramcontent.com/pod-product-compliance
Ingram Content Group UK Ltd.
Pitfield, Milton Keynes, MK11 3LW, UK
UKHW020201200926
475702UK00004B/560
9781032804507